Additional praise for *The Great Equations*

"[*The Great Equations*] puts the face, the person behind the equation to make it that much more important, and also explains the logical and scientific reasons for its importance. . . . Crease dives into the history and application of these . . . equations that give birth to numerous inventions and ideas, and are the basis of our modern world, making *The Great Equations* ever-interesting."

—*Sacramento Book Review*

"[Robert P. Crease] tells the fascinating back stories of 10 equations and shows how modern life is built on them. . . . This book will instill in most nonmathematicians a . . . respect for [mathematics]. . . . It will also give you some strong talking points if you ever run into that math-class smarty-pants again."

—Philip Manning, *Raleigh News and Observer*

"This is more than just a celebration of the great equations. Robert Crease uses his interdisciplinary skills as both a science historian and a philosopher to illustrate that no equation pops into being as the result of a momentary thought of one genius, but is the culmination of years, decades, or even centuries of cultural developments. He also shows how an equation not only affects science and math, but transforms the thinking of all people."

—Dick Teresi, author of *Lost Discoveries: The Roots of Modern Science—from the Babylonians to the Maya*

"Crease makes excellent use of his philosophical background and his long-standing interest in the history of science. . . . [His] writing is intelligent and engaging throughout. . . . Crease's historical understanding is . . . sensitive and well-informed. I admire Crease's deft and ingenious comparisons, which are fresh and striking . . . [and his] insightful and clear exposition, memorable comparisons [and] above all his philosophic thoughtfulness."

—*American Scientist*

ALSO BY ROBERT P. CREASE

The Prism and the Pendulum: The Ten Most Beautiful Experiments in Science

Making Physics: A Biography of Brookhaven National Laboratory

The Play of Nature: Experimentation as Performance

The Second Creation: Makers of the Revolution in Twentieth-Century Physics (with Charles C. Mann)

Peace & War: Reminiscences of a Life on the Frontiers of Science (with Robert Serber)

TRANSLATED BY ROBERT P. CREASE

What Things Do: Philosophical Reflections on Technology, Agency, and Design

American Philosophy of Technology: The Empirical Turn

THE GREAT EQUATIONS

Breakthroughs in Science from Pythagoras to Heisenberg

Robert P. Crease

W. W. Norton & Company
NEW YORK LONDON

Printed in the United States of America
First published as a Norton paperback 2010

For information about permission to reproduce selections from
this book, write to Permissions, W. W. Norton & Company, Inc.,
500 Fifth Avenue, New York, NY 10110

For information about special discounts for bulk purchases,
please contact W. W. Norton Special Sales at
specialsales@wwnorton.com or 800-233-4830

Manufacturing by Courier Westford
Book design by Wesley Gott
Production manager: Julia Druskin

Library of Congress Cataloging-in-Publication Data

Crease, Robert P.
The great equations : breakthroughs in science from
Pythagoras to Heisenberg / Robert P. Crease.
p. cm.
Includes bibliographical references and index.
ISBN 978-0-393-06204-5 (hardcover)
1. Science—Philosophy. 2. Science—History. I. Title.
Q175.C884 2009
509—dc22

2008042494

ISBN 978-0-393-33793-8 pbk.

W. W. Norton & Company, Inc.
500 Fifth Avenue, New York, N.Y. 10110
www.wwnorton.com

W. W. Norton & Company Ltd.
Castle House, 75/76 Wells Street, London W1T 3QT

4 5 6 7 8 9 0

For Stephanie,
beyond category

CONTENTS

The Great Equations

INTRODUCTION

The first equation that most of us learn is a synonym for simplicity:

$$1 + 1 = 2$$

So elementary, yet so powerful! It imparts the very definition of addition: one unit plus one unit equals two units. It is powerful, too, because it exhibits the format for every other equation: in arithmetic, mathematics as a whole, physics, and other branches of science. It shows an arrangement of terms that asserts a particular kind of relationship among them. This little but fundamental equation opens so many doors that it seems like a magic wand. It is virtually *the* entrée into knowledge itself—the first little step, the basis for each of thousands of steps to follow. Richard Harrison, a poet and English teacher at Mount Royal College in Calgary, Canada, once wrote to me of this profound expression:

> $1 + 1 = 2$ is the fairy tale of mathematics, the first equation I taught my son, the first expression of the miraculous power of the mind to change the real world. I remember my son holding up the index finger—the "one finger"—of each hand as he learned the expression, and the moment of wonder, perhaps his first of true philosophical wonder, when he saw that the two fingers, separated by his whole body, could be joined in a sin-

> gle concept in his mind. . . . [W]hen I saw my son's mind open in understanding that "1 + 1" was more than "1 + 1" I saw that small equation as my child's key not to what was wonderful in the outside world, but what was wonderful in him and all of us.

Harrison's description reminds us that learning an equation, at least of the kind as fundamental as 1 + 1, is in effect a kind of journey. It is a journey that takes place in three stages. We begin naïvely without knowing the equation. We are led by schooling or accident or curiosity or intent to comprehend it, often accompanied by dissatisfaction and frustration. Finally, the experience of learning it transforms the way we experience the world, which fills us—naturally, if sometimes only momentarily—with wonder.

This book is about those journeys.

The first human beings lived without equations, and had no need for them. There were no equations in the Garden of Eden, not even on the Tree of Knowledge. None were present in the Sumerian paradise Dilmun, nor in the cosmic egg in which some Chinese believe that P'an Ku hatched the world, nor in any of the other places where various divine creation myths say that the first humans dwelt. Human beings did not even have the *idea* of equations. That idea is a human invention, the result of our efforts to make sense of the world. Even so, human beings did not wake up one day and suddenly decide to invent equations. They acquired the need to over time, and the idea of an equation in the scientific-technical sense first appeared late in human history.

The Latin word *aequare* means to make level or even. Many modern English words spring from this root, including adequate, equanimity, equality, equilibrium, egalitarian, equivalence, and equivocation. The word "equation" at first simply meant a partitioning into equal groupings. The "equator," for instance, is the imaginary line drawn by geographers to separate the earth in two roughly equal halves. Medieval astrologers used the word "equation" to refer to their practice of arbitrarily dividing up the path followed by the sun and planets into equal areas, each allegedly governed by a different constellation.[1]

Meanwhile, numbers and counting were becoming important in human life. Businessmen used them in bookkeeping, finance, and budgets; religious authorities used them for record-keeping of years, seasons, and occasions such as births, deaths, and marriages; and in government officials used them in census, and for surveying and taxes.[2] This generated the need to develop symbols to stand for numbers and quantities.[3] In the third century BC, the Greek mathematician Diophantus took another step, using symbols to stand for *unknown* quantities, and providing some rules for operating on such quantities, including subtraction and addition. He showed not only how to use symbols to describe an unknown number so that it could be determined from known numbers (what is called a determinate equation) but also how symbols could describe something with an infinite set of solutions (a Diophantine or indeterminate equation). It was still a long route to the modern notion of equations. Even Galileo and Newton express their important results—Galileo's law of falling bodies, and Newton's laws of motion—in the form of ratios expressed in words, not in the familiar equation form known to science students. Not until the eighteenth century did natural scientists routinely express their conclusions in the form of equations as we know them today.

A long historical and conceptual journey was required, therefore, to write even the simplest of equations. In 1910, Alfred North Whitehead and Bertrand Russell, two of the greatest mathematicians in history, published the *Principia Mathematica*, a famous, three-volume systematic textbook that derives the foundations of mathematics from the ground up in a purely logical way. When does the equation $1 + 1 = 2$ first make its appearance? Well over halfway through volume one![4]

Thanks to this long journey, the word "equation" eventually came to have a technical meaning as part of a specially constructed language—to refer to a statement that two measurable quantities, or sets of measurable quantities, are the same. (Strictly speaking, then, statements expressing inequalities are not equations.) In this code-like constructed language, indispensable to modern mathematics and science, symbols stand for sets of other things on which various operations (addition, subtraction, multiplication, and division being the simplest) can be carried out.[5]

Ever since this special technical language was developed, each individual equation has had two different types of discoveries. It was originally discovered by the first person to come across it—by the person or persons who introduced it into human culture. And it is rediscovered by each person who learns it since.

The journey to a particular equation has a different kind of setting than that of other historical turning points. The appearance of equations is not framed by bloody battlefields or by clashes of titanic political forces. Equations tend instead to emerge in quiet locations, such as studies and libraries, removed from distractions and encroachments. Maxwell wrote down his world-transforming equations in his study; Heisenberg began to piece together his on an isolated island. Such environments allow scientists to address their dissatisfactions, to explore the gnawing sense that the pieces at hand are not fitting together well and need some adjustment or the addition of something new. Scientists then can focus on some problem that often can be articulated with deceptive simplicity: What is the length of this side of a right triangle? What is the strength of the force between celestial objects? How does electricity move? Can a given pair of seemingly contradictory theories be made to fit together? *Does this make sense?*

When the solution comes, it seems logical and even inevitable. This work is "universally received," writes Roger Cotes, who contributed a Preface to the second edition of Newton's famous masterwork, *Mathematical Principles of Natural Philosophy.*[6] The discoverers often feel as if they've stumbled across something already there. Thus equations seem like treasures, spotted in the rough by some discerning individual, plucked and examined, placed in the grand storehouse of knowledge, passed on from generation to generation. This is so convenient a way to present scientific discovery, and so useful for textbooks, that it can be called the treasure-hunt picture of knowledge. It telescopes a difficult process and leaves us with an inventor, time, and place, and often a cause or purpose. An incident or moment, such as the fall of an apple, becomes a synecdoche that crystallizes the long discovery process. Generations of scholars then earn reputations criticizing the model and complicating this picture. The treasure-hunt picture is useful for everyone!

The treasure-hunt picture of the world, however useful, promotes the view that equations are essential features of the world, not created by human beings. And indeed, we are born into a world that already has equations that "we" did not create. This is why equations sometimes appear to be not really of human origin, around long before we humans got here: On the eighth day, God created the equations, as the blueprint for His work. Or, as Galileo wrote, the Book of Nature is written in mathematical symbols.

But each and every equation had a human genesis. It was put together by a particular person at some specific place and time who felt a need—who was dissatisfied by what was at hand—and who wanted to make sense of things or sometimes merely wanted to make something that appeared hopelessly complicated easier to understand. Sometimes this creative process is buried in antiquity, as is true of the "Pythagorean" theorem, whose principle was known long before Pythagoras. Sometimes the creative process is known in detail thanks to the correspondence, drafts, and notebooks of their inventors, as is the case with equations produced by Newton and Einstein. In each case, however, the equations cannot be said to be their work alone, for these scientists—even when working alone—

were involved in countless dialogues with other scientists in a shared process to make sense of nature.

When British scientist Oliver Heaviside rearranged Maxwell's work into what is essentially their now-famous form—into the form that today is known as "Maxwell's equations"—he remarked that he was simply trying to understand Maxwell's work more clearly. That motivation—sensing that one can express better something that one already knows, but vaguely—might be said of all inventors of equations.

After someone does come up with a new equation about some fundamental issue—when that person has answered his dissatisfaction—it changes both us and the world. Such equations thus do not simply instruct us how to calculate something, adding new tools to the same world, but do something "more," as Harrison put it. In learning 1 + 1 = 2, his son did not merely input a new data point, but became transformed, possessing a new grip on the world. But along with this new grip comes new puzzles, and new dissatisfactions.[7]

Harrison's description, finally, reminds us that equations can inspire wonder. Science is not a robotic activity in which we maneuver in or gaze at the world indifferently, but a form of life with a highly nuanced affective dimension. There is, of course, the celebratory, cork-popping joy that is a natural concomitant to a new discovery or achievement. But if that were the only emotion involved in science—the pleasure of making a discovery that ensures fame and fortune—it would be a sorry profession, for such moments are few and far between. Fortunately, the emotions of science are much more diverse, and thicker, than that. Doing science is accompanied by unfolding feelings at every moment—puzzlement, bafflement, curiosity, desire, the urge to find the answer, boredom that nothing is happening, frustration at getting nowhere, the thrill of being on the right track. Such feelings are always present, not deeply hidden, often overlooked, but easy to notice once we decide to pay attention to them.

When we understand an important equation for the first time, we glimpse deeper structures to the world than we suspected, in a way that reveals a deep connection between the way the world is and how we experience it. At such times, our reaction is not simply, "Yeah, that makes sense," or even what is often called "the Aha! moment." This latter crude characterization goes hand in hand with the treasure-hunt picture of knowledge acquisition, for it simplifies and condenses the emotion of discovery into a single instant. The genuine emotion—wonder—is subtler, richer, and lengthier.

It is natural, though, even for scientists to stop wondering at equations, as they become more wrapped up in the world and their interests in it, and less attentive to the moments of disclosure in which its forms first appear. We lose wonder, indeed, at any instrument or object with which we grow too familiar. Equations can come to seem as just another set of tools that we find lying about in the world, or as onerous chores that we learn out of duty.

Pilots who learn too much about their craft, Mark Twain writes in *Life on the Mississippi*, often undergo a regrettable transformation. As they become increasingly skilled at reading the language of the river, they seem to grow correspondingly less able to appreciate its beauty and poetry. Features of the river—a floating log, a slanting mark on the water, a patch of choppy waves—that once aroused feelings of wonder and awe become increasingly appreciated only instrumentally, in terms of the use they have for piloting. Something similar is true of equations.

But great scientists are often still able to marvel at the breakthroughs of their predecessors. The physicist Frank Wilczek once wrote a series of articles on the simple equation expressing Newton's second law of motion, $F = ma$, calling it "the soul of classical mechanics," and exhibiting toward it the kind of appreciation that is appropriate to souls.[8] The physicist and cosmologist Subrahmanyan Chandrasekhar wrote an entire book on Newton's *Principia*, the book in which Newton proposed his second law of motion, comparing it to Michelangelo's painting on the ceiling of the Sistine Chapel. And a listener to Richard Feynman's famous *Lectures on Physics* can detect throughout his unabashed and spontaneous wonder at the equations he is trying to teach his students. These three Nobel laureates each knew enough to maintain their wonder at the world and at the equations through which we know it.

This book aims to show that there is much more to equations than the simple tools they seem to be. Like other human artifacts, equations have social significance and exert cultural force. This book takes some great equations and provides brief accounts of who discovered them, what dissatisfactions lay behind their discovery, and what the equations say about the nature of our world.

CHAPTER ONE

"The Basis of Civilization":

The Pythagorean Theorem

$$c^2 = a^2 + b^2$$

DESCRIPTION: The square of the length of the hypotenuse of a right triangle is equal to the sum of the squares of the lengths of the other two sides.
DISCOVERER: Unknown
DATE: Unknown

To this day, the theorem of Pythagoras remains the most important single theorem in the whole of mathematics.

—J. Bronowski, *The Ascent of Man*

The original journey to the Pythagorean theorem is forever shrouded in history. But we have countless stories of its rediscovery, both by people who taught it and by people who rediscovered it for themselves. These sometimes have been such powerful experiences as literally to have changed the lives and careers of those who have made them. The power and magic of the Pythagorean theorem arise from the fact that, while it is complex enough that its solution is not apparent at the outset, the proof process is condensed enough to constitute a single experience.

One person whose life it changed was the great political philosopher Thomas Hobbes (1588–1679). Until he was forty, Hobbes was a talented scholar who showed little originality. He was well versed in the humanities but dissatisfied with his erudition. His principal achievement was an elegantly written if sometimes inaccurate translation of the ancient Greek historian Thucydides. He had little exposure to science, despite the exciting recent breakthroughs of Kepler, Galileo, and others, which were then revolutionizing the scholarly world.

One day, while passing through the library of an acquaintance, Hobbes saw a copy of Euclid's *Elements* displayed on a table. This was not unusual: a gentleman who owned a handsome and expensive volume of an important work, such as a Bible, would not store it out of sight but would prominently exhibit it for the benefit of visitors, usually opened to a famous passage or psalm.

Euclid's *Elements* was indeed like a Bible. It set out much of the mathematical wisdom of its time in axioms and postulates; scholars had been analyzing it ever since its appearance in about 300 BC; and its knowledge remained current. No other book at the time, except the Bible, had been as frequently copied or studied. The particular chapter and verse that Hobbes saw was Book I, Proposition 47, the Pythagorean theorem.

Hobbes took a look at the claim: The square described upon the hypotenuse of a right-angled triangle is equal to the sum of the squares described upon the other two sides. He was so astounded that he used a profanity that his acquaintance and first biographer John Aubrey refused to spell out: " 'By G—', Hobbes swore, 'this is impossible!' "[1]

Hobbes read on, intrigued. The demonstration referred him back to other propositions in the same book: Propositions 46, 14, 4, and 41. These referred to still others. Hobbes followed them and was soon convinced that the startling theorem was true.[2]

"This made him in love with geometry," writes Aubrey, adding that Hobbes was a changed man. He started obsessively drawing geometrical figures and writing out calculations on his bedsheets and even on his thigh. He began to devote himself to mathematics, showed some talent—though his abilities remained modest—and embroiled himself in controversies and hopeless mathematical crusades in a manner that still embarrasses his biographers and fans.[3] These episodes are not terribly interesting. What matters is that the theorem transformed him and his scholarship. As one commentator wrote of Hobbes's initial encounter with the Pythagorean theorem, "everything he thought and wrote after that is modified by this happening."[4]

Hobbes began to chastise the moral and political philosophers of the day for their lack of rigor and for being unduly impressed by their predecessors. He compared them unfavorably with mathematicians, who proceeded slowly but surely from "low and humble principles" that everyone understood and accepted. In books such as the *Leviathan*, Hobbes began to reconstruct political philosophy in a similar way, by first establishing clear definitions of terms, and then working out the implications in an orderly fashion. The Pythagorean theorem taught him a new way to reason, and to present the fruits of his reasoning persuasively, in ways that seemed necessary and universal.

Pythagorean Theorem: The Rule

The term, "Pythagorean theorem," is popularly used to refer to two different things: a rule and a proof. The rule is simply a fact. It states an equality between the lengths of the sides of a right triangle: the length of the hypotenuse squared (c^2) is equal to the sum of the squares of the two other sides: ($a^2 + b^2$). That rule has a practical value: it allows us, for instance, to calculate the length of that hypotenuse if we know the lengths of the two sides. The proof is different. It's the demonstration of how we know this fact to be true.

It is confusing that this phrase can refer to both. It's a confusion embedded in the word "theorem." The word can mean a result that is (or is assumed to be) proven. It comes from the Greek for "to look at" or "contemplate," and has the same root as "theater." When people like Hobbes see the Pythagorean theorem, they can pay attention to two very different things: to the product, rule, or thing proven—the hypotenuse rule—or to the process, the proving, or the way it is known.

The rule is extremely important, crucial to describing the space around us. It is invaluable to carpenters, architects, and surveyors in small and large-scale construction projects. This is one reason Freemasons—the esoteric organization said to have been born in medieval stonemason guilds—adopted the Pythagorean theorem as a symbol. One piece of Masonic literature cites the Pythagorean theorem as "containing or representing the truth upon which Masonry is based, and the basis of civilization itself,"[5] and a simplified version of the diagram accompanying Euclid's proof, called the "Classic Form," is often emblazoned on carpets in Masonic lodges. The rule characterizes celestial spaces as well, thus is essential to navigation and astronomy.

This rule was known long before Euclid or even Pythagoras. The fact that sides of specific lengths—3, 4, and 5 units, say, or 6, 8, and 10—create a "set square" with a right triangle between the two shorter sides was an empirical discovery known to ancient craftsmen. Such

A Babylonian cuneiform tablet of about 1800 BC, known as Plimpton 322 after the collection in which it resides at Columbia University. The tablet, evidently a trigonometric table or teaching aid for the rule to figure out hypotenuses of right triangles, contains a table of fifteen rows of Pythagorean triplets.

trios of numbers are called "Pythagorean triplets," and their independent discovery in different lands is not surprising given their simplicity and practical importance. Another ancient discovery seems to have been the rule $c^2 = a^2 + b^2$ for such triplets. A Babylonian cuneiform tablet of about 1800 BC, known as Plimpton 322 after the collection in which it resides at Columbia University, contains a table of fifteen rows of Pythagorean triplets. The tablet was evidently a trigonometric table or teaching aid for the rule to figure out hypotenuses of right triangles. It contains no variables, but it seems to have been intended to communicate the rule via a list of examples.[6]

The rule was also known in ancient India. Applications of it are found in the *Śulbasūtras*, the texts that accompany the Sutras or "sacred teachings" of the Buddha, which seem to have been written between 500 and 100 BC but clearly pass on knowledge of much earlier times. In their instructions for constructing ritual areas they display considerable geometrical knowledge, though it is expressed informally and approximately, and without much justification.[7]

The earliest existing Chinese writing on astronomy and mathematics, the *Zhou Bi Suan Jing* ("Gnomon of the Zhou," containing texts dating from the first century BC but whose contents are said to be centuries earlier), likewise exhibits knowledge of the rule. One application is in a calculation of how far the sun is from the earth. The reasoning process involves a bamboo tube and its shadow, and assumes that the earth is flat; the *Zhou Bi* is famous among historians of science for being "the only rationally based and fully mathematicised account of a flat earth cosmos."[8] The earliest extant version contains an often-reproduced diagram against a

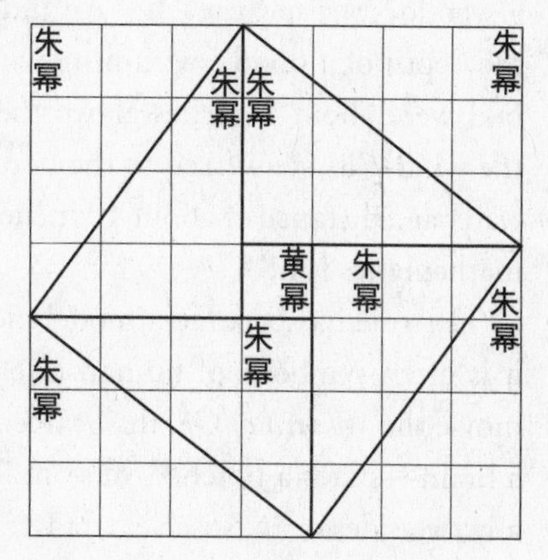

Diagram from a late edition of the *Zhou Bi*. The characters refer to the colors of the squares.

chessboardlike background from which one can readily see that the area of the square built on the hypotenuse is the same as the combination of the areas on the other two sides—but this almost certainly dates from a third century AD source, long after Euclid.

The Babylonian tablet, the Indian *Śulbasūtras*, and the Chinese *Zhou Bi* each exhibit knowledge of the rule as part of a body of mathematical knowledge applied to some other purpose: educational in the case of Plimpton 322, religious in the case of the *Śulbasūtras*, astronomical in the case of the *Zhou Bi*. In these and in other ancient texts the rule is presented without explicit justification, mainly as a way of finding distances and checking results, though occasionally with more formality.

Indeed, the Pythagorean theorem is surely unique among mathematical landmarks for the range of colorful practical illustrations, ranging from prosaic to poetic, over its thousands of years of history, involving the dimensions of fields, canals, clotheslines, footpaths, roads, and aqueducts. From an Egyptian manuscript: "A ladder of 10 cubits has its foot 6 cubits from a wall; how high does it reach?" From a medieval Italian manuscript: "A spear 20 ft. long leans against a tower. If its end is moved out 12 ft., how far up the tower does the spear reach?" An Indian text asks readers to compute the depth of a pond, swimming with red geese, if the tip of a lotus bud were about 9 inches above the water, but was blown over by the wind—its stem fixed to the bottom—and vanished beneath the water at a distance of about 40 inches. These kinds of exercises make mathematics fun!

The rule has become a model piece of knowledge, and knowing it is often symbolic of human intelligence itself. At the end of the movie the *Wizard of Oz*, the Scarecrow—to show he truly does have a brain—states a botched version: "The sum of the square roots of any two sides of an isosceles triangle is equal to the square root of the remaining side." The levity is perfect, for it spares us in the audience from really having to follow it, and keeps what's happening in the realm of fairy tale.

Pythagorean Theorem: The Proof

But proving a rule is much different from just knowing it. A proof demonstrates the general validity of a result based on first principles—for its own sake, not linked with a practical end, and with the focus less on the result than on how one arrives at it; on the process by which we come to trust it. A proof recounts the journey by which we know an equation. To provide the proof of a rule therefore involves a different perspective on mathematics than just stating the rule. For a proof is not an assertion of authority but an acknowledgment of intellectual democracy. It does not simply pass on a piece of wisdom from one's precursors as a tour de force of intellect, a stroke of genius. It does not say, "This is a fact!" or "This is how a genius told us to do it." Instead, the proof of a result says that the journey is something *anyone* can take, in principle at least, thanks to the matrix of mathematical definitions and concepts that we already possess. It therefore says in effect, "Follow this, and you'll see that we know all the steps how to get there *already*!" Giving the proof of a rule therefore establishes a landmark that anyone can get to by following the path indicated, and that one can trust to orient oneself while making further journeys in unexplored territory. Proofs of key equations transform mathematics from a complex terrain into a landscape by erecting landmarks. The rest of mathematics is still present, but in the background.

Although the first proof of the hypotenuse rule is traditionally ascribed to Pythagoras (ca. 569–475 BC), the claim that his proof was the earliest was first advanced half a millennium later, and is almost certainly untrue.[9] The idea of proof seems to have originated in ancient Greece, and took hundreds of years to develop. It culminated in Euclid's *Elements*, which presents mathematical knowledge entirely in the form of explicit, formal proofs. The proof of the Pythagorean theorem is the next-to-last one of Book I. In a right triangle, the square on the side opposite the right angle is equal to the sum of the squares on the other two sides. Proposition 48, the last proof of Book I, is the converse: if the square on one side of a triangle is equal to the sum of the squares on the other two sides, it is a right triangle. The proof is as follows: Build a square on each side of a right triangle. Draw a line from the vertex of the right angle, perpendicular to the hypotenuse, to that square's far side. This divides the big square into rectangles. Each rectangle turns out to be the same size as one of the squares: the sum of the smaller squares thus equaling the area of the square on the hypotenuse. Interestingly, Euclid's proof is associated with the distinctive image created by its lines, and has been called the windmill, peacock, or bridal chair proof after fanciful images that it has been taken to suggest.

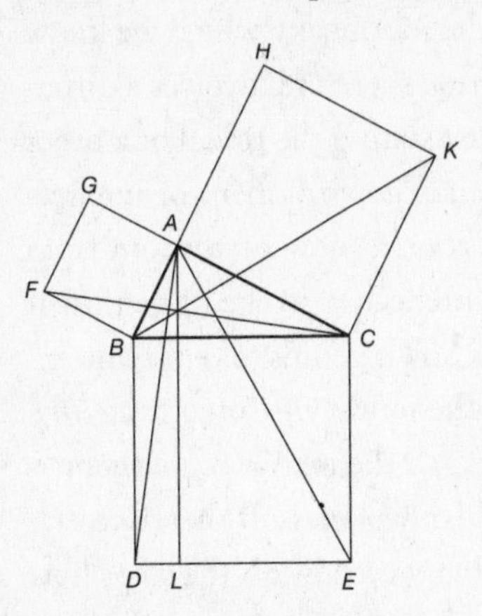

A classic diagram illustrating the proof in Euclid's *Elements*.

Every great discovery seems to generate the irresistible urge to scour through records to see if anyone else discovered it earlier, discovered it but did not write it down, or brushed up against it without discovering it. The Pythagorean theorem, as we seem forever fated to call it, was no exception. For historians, showing how close a people came to proving the Pythagorean theorem appears to be a way to try to show how advanced that civilization was—and claims have been made for the Babylonian, Indian, and Chinese discovery of the Pythagorean theorem based on Plimpton 322, the *Śulbasūtras*, the *Zhou Bi*, and other texts.[10] But in the process, it is easy and tempting to confuse or ignore the difference between the Pythagorean theorem, the empirically determined rule, and the Pythagorean theorem, the proof of the equation.

New Proofs

Occasionally, humans have taken the journey on their own, discovering the Pythagorean theorem without the aid of teachers. One is the French mathematician and philosopher Blaise Pascal, whose father forbade any discussion of mathematics around the household, afraid that the subject might distract his child from the all-important studies of Greek and Latin. But the young Pascal began to explore geometry with the aid of a piece of charcoal, in the process discovering many of the proofs codified in Euclid's *Elements*, including the Pythagorean theorem.[11]

It is also possible to discover new proofs of the theorem. For if the Pythagorean theorem is unique among mathematical landmarks for the range of its applications and examples, it is also unique for the range of ways that it has been proven. Most proofs are based on the same axioms, but follow different paths to the climax. Many—especially the earliest proofs such as Socrates', Euclid's, and in the later Chinese manuscript *Zhou Bi*—are geometrical, where a, b, and c refer to lengths of various sides of shapes, and the proof proceeds by manipulating the shapes and showing something about their areas. Other proofs are algebraic, or based on more complex mathematics where the numbers refer to abstract things, and can even refer to vectors. Some so-called proofs, though, assume results that are proven by the Pythagorean theorem, and so are really circular arguments. The algebraic approach—which the Babylonians understood—is what produced the $c^2 = a^2 + b^2$ version of the rule.

In the fourth century AD, the Greek geometer Pappus of Alexandria discovered a theorem that extended Euclid's. A few centuries later, Arabic mathematician Thābit Ibn Qurra (836–901), working in Baghdad, provided several new proofs in revising an earlier Arabic translation of the *Elements*. Two and a half centuries later, the Hindu mathematician Bhaskara (b. 1114) was so enamored of the visual simplicity of the *Zhou Bi* proof that he redid it in the form of a simple diagram, and instead of an explanation wrote a single word of instruction: "See."

Later, Italian artist Leonardo da Vinci, Dutch scientist Christiaan Huygens, and German philosopher Gottfried Leibniz (1646–1716), all contributed new proofs. So did U.S. Congressman James Garfield, in 1876, before he became the twentieth U.S. president. Indeed, over a dozen *collections* of proofs of the Pythagorean theorem have appeared: in 1778, a list of thirty-eight was published in Paris, in 1880 a monograph appeared in Germany with forty-six proofs, while in 1914 a list of ninety-six proofs was published in Holland. The *American Mathematical Monthly*, the first general-interest mathematical magazine in the U.S., began publishing proofs in its first issues, starting in 1894. With some condescension, it stated that problem solving "is one of the lowest forms of mathematical research," being applied and without scientific merit. Nevertheless, the magazine promised to devote "a due portion of its space to the solution of problems" such as the Pythagorean theorem, to serve an educational purpose. "It [problem solving] is the ladder by which the mind ascends into the higher fields of original research and investigation. Many dormant minds have been aroused into activity through the mastery of a single problem."[12] In 1901, after publishing about a hundred proofs, its editor abandoned the effort, announcing that "there is no limit to the number of proofs—we just had to quit."

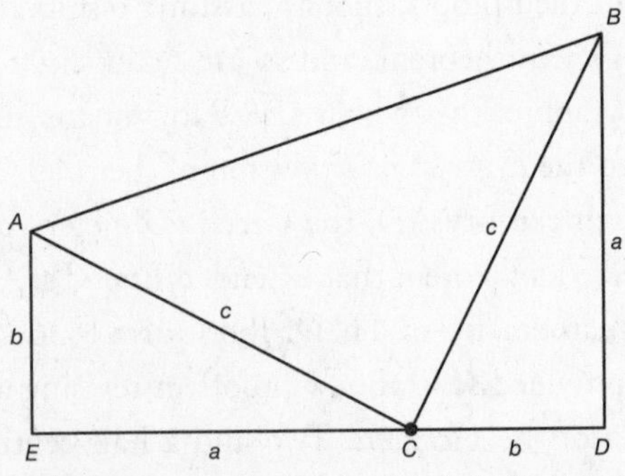

Diagram on the basis of which U.S. President James A. Garfield invented a proof.

One who refused to quit was a schoolteacher and subscriber from Ohio named Elisha S. Loomis—a mason—who had contributed some of the proofs. Loomis continued to collect them, many passed on by teachers of bright youngsters who knew of his interest. In 1927 (by then a college professor) Loomis published *The Pythagorean Proposition*, a book containing 230 proofs; in 1940, the 87-year-old Loomis published a second edition containing 370 proofs.[13] He dedicated both books to his Masonic lodge. Loomis divided the contents into geometric, algebraic, dynamic, and quaternion proofs. Most were geometric: number 31 was Huygens's; 33, Euclid's; 46, da Vinci's; 225, Bhaskara's; 231, Garfield's; and the *Zhou Bi*'s was 243. Of the algebraic proofs, Leibniz's was number 53. Loomis prized the way that the challenge of coming up with a new proof tested the mettle of students and, evidently fascinated by the process of proof, liked to signal interesting proofs, interesting people who had contributed proofs, or to commend youthful contributors.[14] He was disapproving of those who, he thought, disrespected the subject. He chastised some American geometry textbooks that omitted Euclid's proof—possibly to show "originality or independence"—remarking wryly that "the leaving out of Euclid's proof is like the play of Hamlet with Hamlet left out."[15] His final sentence of the second edition: "And the end is not yet."[16]

Loomis was correct; it wasn't. The *Guinness Book of World Records* Web site, under "Most Proofs of Pythagoras's Theorem," recently named a Greek who, it is claimed, has discovered 520 proofs. By the time you read this, more have surely appeared.

Whence the Magic?

All these proofs provoke two questions. The first is: Why isn't one proof enough? We know why one application is not enough: the point of a rule is that it applies to many different circumstances. But proofs? A small number of proofs of the Pythagorean theorem generalize the theorem that Euclid proved, and thus extend what he did.

Most in Loomis's collections, however, are not of that type. Nor do they make the result more certain than it already is. Their fascination lies in the scientific desire not merely to discover, but to view a discovery from as many angles as possible—to convert implicit possibilities, or merely hypothesized or assumed results, into actualities. Science aims to enrich the world, to increase the variety of its forms, to let the reality of the things in the world show themselves. As science progresses, the landscape of the world develops with it.

The second question is: Why all the attention to *this* particular theorem, which has fascinated amateurs and professionals for thousands of years? A part of the answer is surely personal biography: the Pythagorean theorem tends to be the first deep proof that each of us encounters, the first proof where—as Hobbes's experience shows—it is not obvious what it is we are setting out to prove. It is the first journey of mathematical discovery where we find something genuinely new at the other end. But that must be only a small part of the answer, for we also learn other beautiful proofs early on, such as of the irrationality of the square root of two or of the infinity of primes. We also learn proofs that are similar to the Pythagorean theorem (for instance, Proposition 31 of Book VI of Euclid's *Elements*), or much more powerful and useful than the Pythagorean theorem, without these attracting anywhere near the same degree of attention. A striking example of the latter is the law of cosines—$c^2 = a^2 + b^2 - 2ab \cos \theta$—which covers all triangles, not just right triangles, and relates the lengths of the sides to the cosine of one of the angles; the Pythagorean theorem is but a special case of the law of cosines. Yet this law communicates no special magic—partly because one has to know trigonometry to prove it—and one can hardly imagine a Hobbes becoming as transformed.

The full answer as to why the Pythagorean theorem seems magical is threefold: the visibility of the hypotenuse rule's applications, the accessibility of the proof, and the way that actually proving the theorem seems to elevate us to contemplate higher truths and thus acquaint us with the joy of knowing.

First, the theorem characterizes the space around us, and we thus encounter it not only in carpentry and architecture, physics and astronomy, but in nearly every application and profession. The Pythagorean rule for a distance in three-dimensional space—the diagonal of a shoebox, say—is the square root of $x^2 + y^2 + z^2$; for four-dimensional Euclidean space it's the square root of $x^2 + y^2 + z^2 + w^2$; in Minkowski's interpretation of Einstein's special theory of relativity, the four-dimensional space-time version is $x^2 + y^2 + z^2 - (ct)^2$, where c is the speed of light. Suitably adapted, this formula enters into the equations of thermodynamics, in describing the three-dimensional motions of masses of molecules. It also enters into both the special and the general theory of relativity. (In the former, it is used to describe the path of light moving in one reference frame from the point of view of another, while in the latter, in a still more complex extension, it is used to describe the motion of light in curved, four-dimensional space-time.) And it is generalized still further in higher mathematics. In *The Pythagorean Theorem: A 4,000-Year History*, Eli Maor calls the Pythagorean theorem "the most frequently used theorem in all of mathematics."[17] This is not only because of its direct use but also due to what Maor calls "ghosts of the Pythagorean theorem"—the host of other expressions that derive, directly or indirectly, from it. An example is Fermat's famous "last theorem," finally proven in 1994, which asserts that no integers satisfy the equation $a^n = b^n + c^n$ (all variables stand for positive integers) for any n greater than two. Though, being the denial rather than the assertion of an equality, Fermat's last theorem cannot be put in the form of an equation.

Second, as Hobbes's experience indicates, even though the Pythagorean theorem involves a bit of knowledge whose proof seems implausible at the beginning, it can be proved simply and convincingly even without mathematical training. This is one reason why philosophers and scientists from Plato onward use it as an emblematic demonstration of reasoning itself. In *On the World Systems*, Galileo cited Pythagoras's experience proving the theorem to illustrate the distinction between certainty and proof—what we now call the context of discovery and the context of justification.[18] In *The Rules for the Direction of the Human Mind* (Rule XVI), French philosopher and scientist René Descartes used the Pythagorean theorem to show the virtues of symbolic notation, which he was introducing into mathematics. G.W.F. Hegel viewed the proof as "superior to all others" in the way it illustrates what it means for geometry to proceed scientifically, which for him meant showing how an identity contains differences.[19] German philosopher Arthur Schopenhauer, one of the few critics of the way Euclid proved the Pythagorean theorem, viewed that proof as emblematic for another reason. Mocking it as a "mousetrap" proof that lures readers in and then "springs" a trap on them, Schopenhauer thought it logically true but overtly complicated, did not like the fact that not all its steps were intuitive (he much preferred proofs that appealed to intuition), and maintained that Euclid's proof is a classical illustration of a misleading demonstration. Indeed, he saw Euclid's proof of the Pythagorean theorem as emblematic of all that was *wrong* with the philosophy of his day, for it emphasized the triumph of sheer logic over insight and educated intuition. Hegel's philosophical system, for Schopenhauer, was in effect no more than one huge conceptual mousetrap.[20]

Third, the Pythagorean theorem makes accessible the visceral thrill of discovery. Whenever we prove it, we can hardly be said to be "learning" anything, for we learned the hypotenuse rule as schoolchildren. But as the proof proceeds—as we set the problem in a bigger context, and as the little pieces begin to snap together with an awesome inevitability—we seem to be taken out of the here and now to someplace else, a realm of truths far more ancient than we, a place we can reach with a little bit of effort no matter where we are. In that place, this particular right triangle is nothing special; all are the same and we do not have to start the proof all over again to be certain of it. Something lies behind this particular triangle, of which it is but an instance. The experience is comforting, even thrilling, and you do not forget it. The proof arrives as the answer to a puzzle in a language that you did not have beforehand, a language that arrives in that instant yet which, paradoxically, you sense you already possessed. Without that moment of insight, the Pythagorean theorem remains a rule handed down authoritatively, rather than a proof gained insightfully.

The Pythagorean Theorem in Plato's *Meno*

All three components of the magic of the Pythagorean theorem are evident in the earliest known, most celebrated, and most complexly described story of a journey to the Pythagorean theorem. That occurs in Plato's dialogue *Meno*, written about 385 BC, or somewhat more than a century after Pythagoras and almost a century before Euclid's *Elements*. It is the first extended illustration of the mathematical knowledge of ancient Greece that exists. In the *Meno*, Socrates coaxes a slave boy, ignorant of mathematics, to prove a particular instance of the theorem, one involving an isosceles right triangle.

The principal participants are Socrates and Meno, a handsome youth from Thessaly. Meno is impatient, balks at difficult ideas, and likes impressive-sounding answers—a teacher's nightmare. He's been pestering Socrates about how it is possible to learn virtue. Socrates finds it difficult to get Meno's mind going; his name, appropriately, means "stand fast" or "stay put." The word "education" means literally "to lead out." Socrates cannot lead Meno much of anywhere.

At one point, Meno throws up his hands and asks Socrates—in a famous query known as Meno's paradox—how it is possible to learn anything at all. If you don't know what you are looking for, you won't be able to recognize it when you come across it—while if you do know you don't bother to go looking for it. Meno is implying that it is fruitless even to try.

The paradox arises, as philosophers say today, from the mistaken assumption that knowledge comes in disconnected bits and pieces. In reality, we humans notice that something is unknown thanks to the whole matrix of things that we know already. We can extend this matrix—and fill in and flesh out gaps and thin areas—by applying what we know to find what we don't, bringing everything else to bear on it, inevitably uncovering new holes and weaknesses in the process. Acquiring knowledge is not like putting things someone else gives us in a mental warehouse, but a back-and-forth process in which we are constantly moving between parts and wholes, seeing and uncovering new things thanks to what we already know, acquiring a continually expanding base for understanding the world.

This isn't the way Socrates puts it to Meno, of course. Meno cannot digest anything that subtle. Instead, Socrates couches it in a way the gullible lad can relate to, trying to entice him to move. Let me tell you an old legend believed by religious sages, Socrates says. They say souls are immortal, and thus have seen and learned everything under the sun. Deep within us, we already know everything, though during our earthly sojourn we've forgotten just about all. But if we are energetic enough, we can overcome this ignorance by recollecting it.

This legend is Socrates' poetic way of telling Meno that learning is neither like getting something passively handed to you by someone else, nor like automatically following a rule. It's an active and intensely personal process in which you motivate yourself to see something. You have to be on the move. And when you recognize something as true, you see that it belongs in that matrix as if it were a feature already there that you had overlooked. It feels so firmly nestled in your soul that it's like you had it all along. It's as if all the preparation and exercises and proofs you do in trying to learn something serves to help you to unforget it. This is the truth of the myth.

Meno likes the legend. But he still has not gotten the point, and asks for more explanation. Trying another tack, Socrates says he'll show Meno the process in action. He asks Meno to summon one of his slaves—"whichever you like"—and Meno complies. Socrates then coaxes this young slave, a naïve boy innocent of any mathematical training, to go on a little journey, proving the geometrical theorem that the area of the square formed on the diagonal connecting the corners of another square is twice the area of that other square—thus, the

Pythagorean theorem involving an isosceles right triangle. Socrates does so by drawing figures in the sand, step by step, asking Meno to keep him honest by listening carefully to make sure that Socrates does not smuggle any mathematical information into his questions and that the boy is "simply being reminded" and not being spoon-fed.

Modern readers may see what follows as a fraud. They may think that Socrates is pulling the strings, playing with the slave's head, getting the slave just to mouth the words. Modern readers are apt to find the idea of learning as recollection absurd, and think that real learning involves downloading new information into a person's brain to be reinforced with homework and exercises. But if we read Plato carefully, we see that the slave is really learning—learning reduced to its elementals, as Socrates makes sure every new point emerges from the slave's own experience. We see the slave boy going on a little journey in learning the Pythagorean theorem. Out of the infinite number of branching paths to follow, Socrates shows the boy which ones to choose, and provides him with some motivation to choose them.

You know what a square is? Socrates asks the boy, drawing a figure in the sand. A figure with four equal sides, like this? The youth says yes.

Do you know how to double its area? Socrates asks.

Of course, is the reply. You double the length of the sides. Obviously!

That's wrong, of course, but Socrates doesn't let on. A good teacher, he gets the student to spot his own mistake. When he extends the square, doubling the length of each side, the youth sees his error immediately—the new big square contains *four* squares of the original size, not two.

Try again, Socrates says. The boy proposes one and a half times the length of the first side. Socrates draws that square—and the boy sees that he's overshot again.

Socrates asks the slave boy—dramatically, for Meno's benefit—if he knows how to double the area of the square. No, I really don't, says the youth.

This is the key moment! Socrates has, first, gotten the boy to see the limits of his knowledge—what he doesn't know—and, second, dismantled the slave boy's confidence. The boy had assumed that he knew, and now knows that he doesn't know. It's not true, of course, that the slave boy knows nothing. He knows a lot—that the answer to Socrates' question lies in a narrow range, more than one but less than one and a half times the side. But the boy also knows more than he can say, and the knowing without being able to say doesn't feel good to him now. The answer will come, but in a language the boy does not know yet. The slave boy has been made discomforted by encountering something he thinks he should understand and realizing that he does not. That bewilderment provokes a curiosity essential to learning. He is ready to let himself be led—ready to take a journey. It makes him want to *see*. He wants to *move*. We shall encounter the role of something triggering this desire—to move from where one is—again and again in the birth of equations. Sometimes the trigger is a chance event—perhaps the fall of an apple—while at other times it may be a passing remark, puzzling data, or an inconsistency between two theories. Here Socrates has bewildered the boy to induce the boy to want to follow him—a kind of seduction, which was one of the crimes Socrates would soon be accused of and for which he would be condemned to death.

Socrates capitalizes on the youth's bewilderment. He rubs out the drawing and starts again with one of a square, 2 feet on a side, and then puts three other identical squares adjacent to it. Then he adds a new element to the diagram, a line crossing between two opposite corners: "the scholars call it a diagonal." The diagonal is not a totally new element. The slave has seen diagonals in floor mosaics and wall designs (an experience that has already given him an intuition of what is about to happen) and is merely being reminded what one is. But it's new here. The diagonal suddenly casts the problem in a larger, richer context that makes the answer easier to see. It brings about a reformulation of the problem.

Resuming his coaxing, Socrates now easily gets the slave boy to

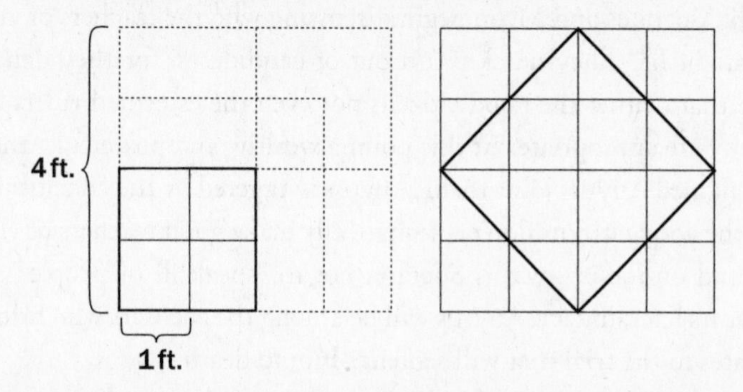

On the left is the first diagram drawn by Socrates. Socrates begins with a square 2 feet on a side and asks the slave boy how to double its size. The slave boy first suggests doubling the length of each side to 4 feet, but this quadruples the area of the square; then increasing the length of the side by one-and-a-half times (by 1 foot), but this, too, increases the area by too much, to 9 feet. In the second diagram on the right, Socrates introduces diagonals, and the slave boy then realizes that the area of the square enclosed by them is twice that of the original square.

---

see that a square built on that diagonal is equal to twice the area of the first square.

Socrates turns to Meno, and tries to lead him on a journey of another kind. Has the boy gone from not knowing to knowing? Yes, admits Meno. Has Socrates fed him any information? No. He's found the answers within himself? Yes. While these freshly stirred up opinions, being new, are "dreamlike" now, Socrates continues, with more questioning—to make sure the learning is secure and does not slip away—the slave boy will carry this knowledge around inside him, and his "knowledge about these things would be as accurate as anyone's." (We call such additional questioning "homework and exercises.") And if we insist on sticking to the terms of Meno's paradox and say that the boy either knew or didn't know, then he must have known but forgotten, just as the legend said. Right, Meno admits. I wouldn't swear to all of the legend, Socrates says, but I'm sure it's got grains of truth.

Now that Meno is satisfied that learning is possible, the conversation reverts to the original question of virtue and how it might be taught. Socrates and Meno begin discussing who the teachers of virtue might be. They quickly run out of candidates, for they determine that neither the good citizens nor even the esteemed rulers of the city are appropriate. At this point a wealthy and powerful Athenian named Anytus joins them. Anytus is angered by the conclusion that the good citizens don't automatically make good teachers of virtue, and ominously warns Socrates not to "speak ill of people." A few years later, in fact, Anytus will be among the accusers who bring Socrates to the trial that will sentence him to death.

In the play-within-the-play, we see a lot of things; we readers go on a journey as well. We see the Pythagorean theorem taking shape before our eyes. We see the slave boy take a journey in learning the theorem. We see that Socrates is leading the boy, but also that a condition for being led is that the slave boy moves himself. We see Meno going on a journey, looking at the boy moving from ignorance to knowledge. We see what knowing is like: when we get stuck, we can go forward by adding elements to enrich our matrix of terms. The new line—the diagonal—is not present at first. Once introduced, it is as observable as any other line and enriches the matrix of elements to make the path clear. A more sophisticated and concise picture of education in action has never been penned.

But Plato is also showing us, the readers, something about our own situation. The play-within-the-play shows us that we are in the position of the slave boy without the benefit of Socrates to ask us the right questions and give us the right new terms. To some extent, human situations inspire their own implicit questions and create their own uncomfortable feelings, and chance sometimes drops in the diagonal for us; still, the answers often come in a language we don't know yet, and we will have to forge ahead and create a denser language on our own. Like the young Pascal, we will have to learn how to add that next diagonal ourselves. Plato is also telling us to keep asking questions. Human beings are always tempted to turn what they know into something fixed and congealed, always exposed to the danger of having their deepest truths turn into illusions, reality into dreams. That is why Socrates famously denounced books in the *Phaedrus*, calling them "orphaned remainders of living speech," which don't talk back. The only way out is to keep questioning, keep interrogating our experience, keep moving.

Plato has one final trick up his sleeve. He is using the episode to point out that, in our efforts, we will encounter two serious dangers. One is inertia from lazy academics, modern Menos, who will insist that we cannot really learn—that all we can do is add something that looks like what we already have and even if it looks new it's only a projection, a construction. The second danger is from our politicians and their henchmen, modern Anytus's, who will tell us that patriotism and the faith of the rulers takes precedence over scientific inquiry. Each group seeks to deny human cultural achievement in a different way. We will have to be patient with the first group; careful, even obsequious, with the second. In one of the most intricately plotted short pieces of literature extant, Plato uses the episode of the Pythagorean theorem in the *Meno* to show us that the journey of truth is much more difficult and perilous than the comfortable quest it is generally billed to be.

INTERLUDE

Rules, Proofs, and the Magic of Mathematics

We all know the rule, but do we all know the proof? The Pythagorean theorem can be proven in many ways, sometimes even in ways that do not involve a single word. The Cité des Sciences et de l'Industrie in Paris, the largest science museum in Europe, has a visual wall display of the theorem in three dimensions. Three solid but hollow figures are built, one on each side of a right triangle, partly filled with colored liquid that can flow from one solid into the others. When the display revolves, the liquid completely fills the solid built on the hypotenuse with no remainder—but then flows into the other two solids, filling them without remainder! And a nineteenth-century edition of Euclid's *Elements*—known as "one of the oddest and most beautiful books of the century," was exhibited at the famous Crystal Palace exhibition in London in 1851, and today regularly sells on eBay for thousands of dollars—cleverly used colored lines and figures to condense most of the text of the proofs, including that of the Pythagorean theorem, into almost purely visual presentations.[1]

Philosopher David Socher has a clever way to demonstrate the difference between the Pythagorean theorem, the rule, and the Pythagorean theorem, the proof, to his students of all ages.[2] Without telling them what he is up to, he hands each a large white square and four colored triangles. "I simply explain that we're going to do a little demonstration. I'm going to ask you to move the pieces in certain ways. It's not any kind of trick. It's not hard and it's not a speed test. It's a friendly little demonstration." He then asks them to arrange the four triangles (each of which happens to be 3 inches × 4 inches × 5 inches) on the square (with 7-inch sides) in two different arrangements. The students readily agree that, in each case, there is the same amount of white space left over. He asks what this says about triangles, and the students usually don't say much. He asks what they know about triangles, and at least one person usually repeats the Pythagorean theorem, without realizing the connection to what's in front of them. "*Pay dirt*," Socher writes. For with a word or two more, the connection between rule and proof suddenly descends.

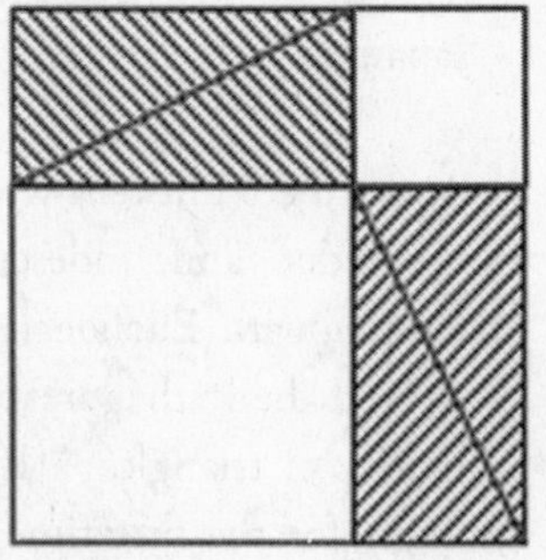

Four triangles in two different arrangements.

That is the unforgettable moment, the kind that we remember—and even long for—as adults. In *Quartered Safe Out Here: A Recollection of the War in Burma*, the British novelist George MacDonald Fraser tells of showing the Pythagorean theorem to his comrade Duke one night after their regiment had dug in along the road to Rangoon during World War II. Tired of conversation about cigarettes, the war, and the Japanese, Duke is on edge; later that night he will die horribly after a series of accidents and misunderstandings, cut almost in two by a line of friendly machine gun fire as he stumbles in the dark. He asks Fraser to tell him "something educated," wanting "a minute's civilised conversation in which every other word isn't 'fook'." Fraser offers to "prove Pythagoras," and Duke, delighted, promptly bets him he can't.

> I did it with a bayonet, on the earth beside my pit—which may have been how Pythagoras himself did it originally, for all I know. I went wrong once, having forgotten where to drop the perpendicular, but in the end there it was, and the Duke's satisfaction was such that I went on, flown with success, to prove that an angle at the center of a circle is twice an angle at the circumference. He followed it so intently that I felt slightly worried; after all, it's hardly normal to be utterly absorbed in triangles and circles when the surrounding night may be stiff with Japanese.[3]

And Albert Einstein wrote in an autobiographical essay of the "wonder" and "indescribable impression" left by his first encounter with Euclidean plane geometry as a child, when he proved the Pythagorean theorem for himself based on the similarity of triangles. "[F]or anyone who experiences [these feelings] for the first time," Einstein wrote, "it is marvelous enough that man is capable at all to reach such a degree of certainty and purity in pure thinking."[4]

Einstein's experience shows yet another kind of thrill that the Pythagorean theorem can teach. For those who do not merely learn how to prove it, but manage to come up with a new proof, the experience teaches the thrill of creativity itself. The person who does this is not merely watching the proof come into being as a spectator watches the unfolding of a little play—that person has become a playwright, doing what mathematicians do, practicing mathematics as a creative art, experiencing the joy of creation, discovering that the true essence of mathematics is doing more mathematics. Such a person has discovered the power of discovery.

For Plato, Hobbes, Descartes, Hegel, Schopenhauer, Loomis, Einstein, Fraser, and countless others, the Pythagorean theorem was far more than a means to compute the length of hypotenuses. To someone who follows the reasoning, something more than the bare result becomes evident. In the experience of one thing—the content, the mathematics—there is a moment of manifestation in which something else, a structure of reasoning, also comes to appearance. It is a rugged, hardy, stubborn piece of knowledge that no religious conviction can dispel, no political ideology can disguise, no academic artifice can conceal.

In a similar way that $1 + 1 = 2$ imparts the idea of addition, so the Pythagorean theorem imparts the idea of proof making. It makes possible what philosophers call categorical intuition: one can see in it more than bare content, but a structure of the understanding. It involves a journey short enough that its stages can be taken in at a glance to illustrate the journey of knowledge itself. It is a proof that demonstrates Proof.

CHAPTER TWO

"The Soul of Classical Mechanics": Newton's Second Law of Motion

$$F = ma$$

[Newton's own] DESCRIPTION: A change in motion is proportional to the motive force impressed and takes place along the straight line in which that force is impressed.
DISCOVERER: Isaac Newton
DATE: 1684–87

Newton's second law of motion, $F = ma$, is the soul of classical mechanics.

—Frank Wilczek, *Physics Today*

The equation $F = ma$ is shorthand for Newton's second law of motion. It is the $1 + 1 = 2$ of classical mechanics. It seems obvious and straightforward. The equation appears simply to translate an ordinary experience into measurable terms: push something and it either starts to move or moves differently.

Yet, like $1 + 1 = 2$, $F = ma$ erupts into mystery when looked at closely. It does not, in fact, refer to ordinary experience but to an abstract world of zero resistance: in the real world we have to continue pushing things like desks and carts to keep them moving at the same speed. The equation does not incorporate Einstein's famous discovery of the interchangeability of mass and energy. It gives center stage to force—a concept absent from most formulations of contemporary theories like relativity and quantum mechanics. Finally, the equation seems, in a contradictory way, to be both a name and a description. It seems both to define force, mass, and motion, and to state an empirically discovered and testable relationship among them.

How can such an elementary equation about something as ordinary as motion conceal so many complexities? The answer can be gleaned in the remarkable historical journey that led from ancient times to the equation's formulation in the seventeenth century. To arrive at this equation, human beings had to train themselves to look at motion in new ways—to learn to look at different aspects of it, and to change how they thought about what they saw. In the course of this vast journey, new sights slowly and progressively came into view, occupied center stage, and then vanished off the horizon, with each familiar landscape slowly yielding to another, until the travelers found themselves in an entirely new world.

Greek Notions of Motion and Change

The journey began in primitive times, when human beings saw the world as ruled by deities. This was natural and inevitable, perhaps the simplest and most straightforward way to make sense of things. All humans acquire a notion of force from individual experiences of pushes and pulls of daily life, in applying our muscles to lift, squeeze, or roll things. Generalizing that experience, early humans could readily conceive everything in nature—from nearby phenomena like thunder and rain to the movements of distant bodies like the sun and stars—as the result of spirits behaving and misbehaving, exerting their particular internal forces. Thus early ideas of force were closely connected with religious ideas of the direct presence of gods in the world.[1]

The earliest humans naturally tried to control nature by pleasing the spirits through ritual and prayer—the earliest forms of technology. But that did not succeed in bringing about the desired control. A far more effective way to predict and influence nature seemed to be to pay attention to the kinds and amounts of changes in nature—the recurrence of the seasons, the various movements of the planets and stars, the behavior of fire and floods, and so forth. But nature is so varied! Sunlight and clouds, tides and storms, plants and animals, men and women, plans and ideas, houses and cities are constantly being born and dying, rising and falling, changing colors and forms, and moving about. How could one make sense of all these motions?

The Greek philosopher Aristotle (384–322 BC) was the earliest we know who drew up a systematic account of all kinds of motion or change—he used the same word, *kinesis*, for both. Kinesis is so important, he thought, that to understand it is tantamount to understanding nature itself, and he created a framework to include all varieties of kinesis: of animate and inanimate objects, with and without human intervention, on earth and in the heavens. He distinguished several kinds of kinesis: the substantial change of a thing being born or dying (fire consuming a log); the quantitative change of a thing growing or shrinking; the transformational change of one property changing into another (a green leaf turning brown); and local motion, or something changing its place.

Aristotle viewed these changes with biologically trained eyes. He regarded the world as a kind of cosmic ecosystem that contained many different levels of organization. Motion in this ecosystem is almost never random or chaotic, but a process of passing from one state to another in which something existing only in potential (a formal principle) is underway to being actualized. Many levels of organization are built on top of each other—human beings make up a state, organs make up a human being—so that any event is shaped by a complex network of different kinds of causes.

Aristotle understood this cosmic ecosystem in the framework of a set of key distinctions. He distinguished, for instance, between two kinds of motion: natural, or violent and forced. Natural motion was that of things moving themselves in their own habitats—acorns growing into oaks, or eggs into chickens—where the change actualizes some innate principle in the substance itself. Forced or violent motions occur when the change is imposed from the outside, as what happens to oak trees when humans fell them to build houses, or to chickens when humans slaughter them for food.

Aristotle also thought it mattered where a change happens. In the earthly realm below the moon, substances are composed of different mixtures of earth, air, fire, and water, and objects don't move constantly but intermittently. In the heavenly realm, objects are made of an unchanging substance called "ether," and move ceaselessly and circularly. If today we find this unjustified, it is a sign of how far we have traveled since Aristotle's time and how our sight has changed, for his ideas were based on rational argument, logical deduction, and careful observation. For hundreds of years, astronomers in Greece and elsewhere had never witnessed any changes in celestial behavior, nor seen anything but circular motion.[2] Only circular movement can continue unceasingly, he thought, and only some special substance, not known on earth (hence the strange name ether), does not experience change.[3] In the celestial realm, motion is initiated by a so-called unmoved mover, which drew the celestial spheres into motion. This was Aristotle's analogue to God, though it was impersonal and not something with which one could have what we twenty-first-century humans call a "relationship." The celestial spheres, by various intermediaries, transmit the motion to the terrestrial sphere. Thus all motions in the cosmic ecosystem, however tiny, are connected, in a mediated way, with the first principle of the universe, and ultimately have to be understood in that context.

When Aristotle discussed what we call motion, then, we onlookers from 2,500 years later have to be careful not to read in our own assumptions. When he speaks of local motions, it is generally in the context of events such as a horse pulling a cart on the road or shipbuilders pushing a boat. Such events arise from a complicated network of purposes, plans, and designs that are being actualized, of which the local motion is only one aspect. And when Aristotle does discuss that aspect, he is not propounding and defending some hypothesis about local motion apart from the event itself, but rather speaking in general terms about the work required to accomplish such tasks to illustrate some other point. In these kinds of events, furthermore, the role of acceleration is almost nil, and rules of thumb such as "A force that moves a body against a certain resistance a certain distance in a certain time moves the same body half the distance in half the time" work fine. Though at one point, in an assertion to become infamous two millennia later as the target of falling-body experiments, he remarked that "If half the weight moves the distance in a given time, its double (i.e., the whole weight) will take half the time."[4]

It is difficult for us to see the world as Aristotle did. Our thoroughly quantitative understanding of motion has become second nature, thanks to familiar concepts like uniform speed and acceleration, to a technologically rich environment containing instruments like digital clocks and speedometers, and to our practical experience with equipment that depends on such concepts and instruments. The experience of Aristotle and his contemporaries was much different. They had neither the experimental instruments nor a mathematical framework for measuring and analyzing motion, and no urgent reason to seek them. They found it plausible to understand movement in terms of form and purpose, not of how quickly the motion takes place.

Aristotle and his contemporaries were not familiar with any of the key components of $F = ma$. His notion of speed or "quickness" was simply that some things cover more ground in the same time than other things—what we would call average speed or overall speed, rather than instantaneous speed, or speed at a particular instant.[5] His notion of acceleration was simply that some things go more quickly as they approach their natural place.[6] He had no concept of mass: of a resistance to being pushed that is not identical to weight. And he had no quantitative notion of *dynamis*, the capacity for motion, nor any units to measure it in.

Nevertheless, it made sense to view nature as a vast ecosystem, comprised of different types of substances acting through different kinds of inner compulsions on other substances, affecting others and being affected in turn, everything with a different purpose to play, all essential to the maintenance of the ecosystem with its qualitatively different domains. Understanding nature required seeing its phenomena in their perfected state—"perfected" in the sense of fully deployed or actualized (the adult tree, the mature human being, the well-functioning society), the phenomena having attained their telos, or end, for in that condition the whys and hows of phenomena are most clear.

Aristotle liked to say that the wise person seeks only as much exactitude as the subject matter allows. He described what he saw, to the most appropriate level of precision that he could. What appeared to matter in understanding the motions of nature was the role that things like form, matter, and purpose play in converting potentiality into actuality. And these ultimately referred to the unmoved mover, who communicates through love via the outer spheres to the moon and then to the sublunary world.

Steps Beyond Aristotle

Aristotle's picture of nature had an enormous impact on Western civilization. His ideas were passed on by students at the Lyceum, the school he founded, and by commentators on his works—at first Greeks, and then, from the ninth to the twelfth centuries, Arabs, from whom later Western scholars learned about Aristotle.

But aspects of Aristotle's picture were not completely satisfying, not even to him. He seemed puzzled, for instance, by how things such as projectiles and potters' wheels moved after the initial push. If a mover has to be in constant contact with what it moves, why doesn't a stone or arrow plunge to the ground after leaving the hand or bow? Aristotle considered two possibilities. One was that the mover (thrower or bow) impregnates or impresses a force on the medium (air) around the projectile (stone or arrow), which then keeps the object in motion.[7] The other explanation, the doctrine of antiperistasis, was that air displaced in front of the projectile rushes around to the back to squeeze the projectile forward.[8] Aristotle was not comfortable with either explanation.

Later thinkers, too, were dissatisfied by this and by other elements of Aristotle's account of motion. Some objections were logical, some empirical, some both. The result was discussion, inquiry, modification of Aristotle's concepts, the introduction of new concepts, and—during a journey of thousands of years—a slow shift of attention to different aspects of motion that would lead, eventually, to $F = ma$. We will travel a long way without seeing anything that resembles its components. But each step of the journey was essential. What follows are some of the steps.

In the third century BC, Strato (340–268 BC), a Greek from Lampsacus in Asia Minor who took over as head of the Lyceum in 287, developed and extended Aristotle's thinking in an influential book called *On Motion*. Strato found he had to revise or even reject some of Aristotle's ideas to make them consistent with logic or experience. One was the idea that there were two kinds of natural movement: up and down. Strato argued that all things naturally go down toward the earth's center, and that if light things like fire and smoke rise, it's because they are displaced or "squeezed out" by heavier stuff. Strato was also bothered by two observations that seemed to suggest that things pick up speed as they fall. One was that when rainwater pours off a roof, the flow is continuous at first but then breaks into droplets, which could not happen if the water weren't moving more quickly.[9] The other was that, when you drop a stone to the ground from high up, the impact is more powerful than when you drop it from just above the ground. How could this be? The stone hasn't gotten heavier! It must have picked up speed, Strato concluded, meaning that a falling body "completes the last part of its trajectory in the shortest time," a rudimentary notion of acceleration more sophisticated than Aristotle's.

In the sixth century AD, John Philoponus ("Lover of Hard Work," ca. 490–570) further revised Aristotle's ideas on motion. Philoponus argued on logical grounds that motion was possible in a vacuum (something Aristotle had rejected), and solved the problem of what happens when force equals resistance by declaring that speed is determined by an excess of force over resistance. Philoponus is also the first person known to have actually experimented with falling bodies of different weights, discovering, as Galileo would a thousand years later, that they fall at approximately the same rate. But Philoponus's most original and far-reaching revision of Aristotle's ideas concerned projectile motion. He rejected antiperistasis; if the mover communicates motion to the air behind the projectile, why can't we send stones and arrows flying by stirring up the air behind them with our hands? Philoponus proposed that, when we throw a stone, our hand impresses a force not on the air but on the stone itself. This "impressed force" causes the motion to continue from *inside* the projectile, but is slowly consumed in overcoming the resistance of the medium and the natural downward force, and eventually used up as the natural motion takes over or the stone hits the ground. This view was still faithful to Aristotle's in that it assumed that an object did not move by itself but always required contact with some other cause, such as the weight of a falling body or the impressed force borrowed from the hand. What was new was that this cause could be internal, not external, to the moving body. This idea led Philoponus and his followers to see the world differently. They no longer needed to distinguish natural and enforced motions, nor to separate the earthly and heavenly realms. God created the heavens, and then used impressed force to keep them going, there being no medium in the heavens to exhaust them. Philoponus's influence helped to inspire scholars trying to understand motion to shift their attention from its end point—the goal or purpose of the motion, whether on earth or in heaven—to its beginning point, or what set it in motion.

Philoponus's modifications, especially concerning impressed

force, influenced Islamic commentators on Aristotle, such as the Spanish Islamic theologian Ibn Bājja (known to the West as Avempace, ca. 1095–1138), the Spanish Islamic theologian Ibn Rushd (Averroës, who objected to Philoponus's view, 1126–1198), and the Persian Islamic theologian Ibn Sīnā (Avicenna, 980–1037). The latter translated Philoponus's idea of impressed force into Arabic as *mail qasrī* (violent inclination). Heavier bodies can retain more *mail* than light bodies, which is why you can throw a stone farther than a blade of grass or a feather. And the Arab commentators concocted other situations where Aristotelian explanations were dissatisfying: What would happen if a tunnel were dug through the earth and a stone dropped in it? Would a thread attached to an arrow's tip be pushed forward? In Ibn Sīnā's work, even more clearly than in Philoponus's, the key to motion is to be found not in formal and final causes but in efficient and material causes.

This shift in attention is clear in the work of John Buridan (ca. 1300–1358). Further developing the ideas of Philoponus and Ibn Sīnā, Buridan gave impressed force the technical name *impetus*, by which it would be known until modern times. Unlike impressed force, impetus did not use itself up but was permanent; a body could only lose it by transferring it to something else.[10] Impetus may sound like our notion of inertia, but unlike inertia it was still a cause. Thus the new framework was still Aristotelian, for it retained the distinction between natural and violent motion and viewed a projectile such as a stone or arrow as continually moved by the action of a cause, though this cause (the impetus) functioned within, not as per Aristotle without, the body. But several key Aristotelian puzzles—such as the question of projectile motion and how bodies fall—had vanished, for the thrower transmits impetus to the stone rather than to the medium, while a falling body acquires impetus as it falls, explaining why it picks up speed. The idea of impetus helped produce a primitive notion of mass—of resistance in a body different from weight—because things like cannonballs can "hold" more impetus than light wood. And it explained why the celestial spheres move forever without divine intervention: with no resistance in the heavens, the spheres need no intervention. God created the spheres, and then gave them impetus, which is why God could rest on the seventh day without His creation grinding to a halt. God's role is thus reversed from the way Aristotle saw it: God is not the continually active final cause that draws the spheres into motion and toward which they strive, but the efficient cause that sets them in motion in the first place.[11]

For the next 300 years, scholars used the idea of impetus to understand and explain motion. It reduced, but did not eliminate, the need to make qualitative differences between natural and violent motions, different kinds of substances, and the heavens and the earth. It also allowed for the development of new conceptions of force, such as percussive force (something that could act once, such as a bat striking a ball) and force that acts continually from a distance, such as whatever pulls objects to earth. It facilitated the development of the idea of mass; some internal density of matter in a body that resists force that is related but not identical to the body's weight. Scholars were beginning to look at motion all by itself—in what philosopher Charles Taylor calls the "immanent frame"—and to study it in certain (but not all) respects without reference to the purposes, plans, and designs of the rest of the universe. They begin to see what we would call a separation between physics and metaphysics. The scientific world and the lived world were beginning to come apart.

Mathematics, meanwhile, was being applied to the world in new ways. Numbers had been used in human affairs for centuries, of course, but scholars were developing new tools to deepen and extend their use. One was Thomas Bradwardine (ca. 1300–1349), from Merton College at Oxford University, later the Archbishop of Canterbury, so renowned in his time that he is mentioned (briefly) in Chaucer's *Canterbury Tales*. Bradwardine developed the foundations of a mathematical framework able to handle velocity, instantaneous velocity (velocity at any particular instant, as opposed to average velocity over a time interval), uniform velocity, uniform acceleration, and changing acceleration.[12] He recast the views of Aristotle, Philoponus, and Averroës in mathematical form, displayed their limitations, and stated his own law. Bradwardine's work was further developed by Nicholas Oresme, who showed how numbers could be applied to describe any continuously varying quantity, such as movement, heat, and so forth. You "pretend," Oresme said, that you are measuring a geometrical surface.[13]

Bradwardine, his followers (known as the Oxford "calculators"), Oresme, and others of the time were not experimenters, but produced a mathematically sophisticated framework for later experimenters. Their work paved the way for the widespread application of numbers to the world by people who saw no need to pretend when using them. In the late sixteenth and early seventeenth centuries, a vast extension of numbers into the world took place in which new kinds of phenomena were quantified. William Harvey (1578–1657) quantified how the heart pumped blood; Santorio Santorio (1561–1636) quantified the intake and excretion of food by the body.[14] Such quantification profoundly affected how motion was understood. Many earlier thinkers, such as St. Thomas Aquinas, had understood many kinds of change as happening through the increased or decreased participation of a body (an apple, a person) in the form of something else (redness, goodness). But the increasing mathematization encouraged the view of all change as taking place through addition or subtraction, similar to the way a line segment changes length by adding or subtracting a segment of a discrete length.

Other events introduced new dissatisfactions into what remained of Aristotle's framework, further paving the way for $F = ma$. A supernova occurred in 1572, another in 1604, and astronomers were able to show that these events occurred not near the earth but in the celestial realm; evidently, things there change just as down here. In 1609, Galileo Galilei (1564–1642) used a telescope to bring the heavens closer, suggesting more similarities with this world than suspected. Such events fostered attempts to develop a physics for the entire universe. Other developments changed the way humans looked at forces. In 1600, the physician to Queen Elizabeth, William Gilbert, wrote a work on magnetism—one of the first treatises of modern science—arguing that magnets work by emitting rays. Indeed, Gilbert said, the earth itself is a magnet, emitting a force that extends in space and varies in strength with distance. This helped promote the idea of a force that could act, in a distinctly un-Aristotelian matter, without contact. Johannes Kepler (1571–1630) published two books, *New Astronomy* (1609) and *Harmonies of the World* (1619), that provided three mathematical laws governing planetary orbits, a kind of mathematical script in the world. Kepler argued that God could have caused the planets to move any way He wished, but decided to have them obey mathematical laws *because* He found such laws beautiful. The mathematics of the world was the script of the world, and its final cause as well.

Galileo was more radical: not only *can* we read the mathematical script of the world, but we should *only* do so and forget other kinds of causes. The "book of nature," he wrote, is "written in mathematical figures." Seeking fantasies such as final causes is not worthwhile. To help read this book, Galileo introduced a brilliant thought experiment: Imagine what would happen on a plane of infinite extent and no resistances, and try to understand how things would move on it, and he proceeded to investigate by staging experiments with things like swinging pendulums and balls rolling down inclined planes. This involved treating space and time quite differently from the way Aristotle had. While Aristotle had treated space as a boundary, Galileo saw it as a container with geometrical properties. To understand motion, you look at how many units of space (Galileo measured it in cubits) an object covered in how many units of time (pulse-beats or water drops). In the process, Galileo discovered the famous law of motion of a falling body—stated by him as a ratio, though nowadays we always state it as the equation $d = at^2/2$, rewriting Galileo in our terms the same way Bradwardine did his precursors. This was the first true mathematical law of nature, the first piece of science to be written in the same language that $F = ma$ would be. Galileo was also able to analyze motions by such things as cannonballs, marbles, and pendulums into two components: a uniformly moving one (push sideways) and an accelerated one (downward).

But Galileo did not yet have the components of $F = ma$. He was still in the shadow of the Aristotelian tradition that distinguished between the natural tendencies of a body, such as free fall, and "violent" pushes or pulls applied from the outside—and tended to think of force in terms of the latter. He did not, for instance, think of a falling body as accelerated by a force, which inhibited him from arriving at a general conception of force and its role in motion. Compounding this was a terminological uncertainty; Galileo was unsure about what to call force, and often uses nearly synonymous terms such as impetus, moment, energy, and force (from the Latin *fortis*, for strong or powerful).[15] When he spoke of force, it was generally not what we call continuous force but instantaneous force (one thing striking another, like a billiard cue a ball or a hammer a nail), or a series of them added together. And Galileo had but a dim recognition of mass—a property of bodies that resists a force, a density of matter related to weight but not identical to it, present even in the absence of gravity. Many of Galileo's ideas, indeed, sound strange to modern ears, as his remark that circular motion is proper to ordered arrangements of parts, or that straight-line motion indicates that a thing is out of its natural state and returning to it. Science historian Richard Westfall calls Galileo's conception of nature "an impossible amalgam of incompatible elements, born of the mutually contradictory world views between which he stood poised."[16]

Galileo Galilei (1564–1642)

Newton

But all these elements appear clearly and systematically in the *Principia* (1687) of Isaac Newton (1642–1727). Newton had learned much from Galileo and other precursors, and developed a generalized and truly quantitative conception of force both continuous and instantaneous, relating it to quantitative changes in the motion of bodies. In the *Principia*, changes in motion are not explained by what is inside them, but only by the forces that befall them from without. This was a new way of looking at motion—not at its why, but exclusively at its how.

Isaac Newton (1642–1727)

This was a modest step beyond Galileo and other close contemporaries of Newton such as Leibniz and Descartes, but thanks to its significance was monumental. It changed the ontology of nature—the way we conceive the most basic units of explanation of the reality we see. Most of Newton's contemporaries conceived of these units as the bodies themselves, which affected other bodies through various mechanisms that brought about different kinds of changes. Newton transformed that, asserting that explanations of motion had to be in terms of the forces that changed the motion of a mass. The three basic terms in the ontology of motion were now *force*, *mass* (what resisted force), and *acceleration* (change in motion). And each of these was quantitative, measurable.

The *Principia*, the most revolutionary single publication in science, much like Euclid's *Elements*, lays out its contents as if they were deductions from self-evident axioms. It contains three Books (the first two called "The Motion of Bodies," the third "The System of the World") preceded by a Preface, eight Definitions, and a set of "Axioms, or Laws of Motion." In the Definitions, we see Newton, sometimes clumsily, developing the components that would be included in $F = ma$—especially the idea of force—from out of the ideas of his predecessors. Definition One is of mass, or quantity of matter; Definition Two of quantity of motion. Of the following Definitions, of force of various kinds, some court ambiguity and even confusion: Definition Three, for instance, is supposedly about "inherent force," but is actually about what we call inertia, describing it in terms of impulse. This definition is best viewed, one commentator says, as "a concession to pre-Galilean mechanics."[17] Definition Four is the key one, and defines "impressed force" in more modern terms as "an action exerted on a body to change its state either of resting or of moving uniformly straight forward." Newton thereby generalized the notion of force developed by his precursors, extending it from instantaneous forces to continuous forces, which was a product of Newton's intuition, science historian I. Bernard Cohen notes, because it was "a step which he never justified by rigorous logic or by experiment."[18] And as Newton writes later, "this concept is purely mathematical, for I am not now considering the physical causes and sites of forces."[19] The remaining four definitions are of what amount to other aspects of force. In a commentary, Newton warns that "although time, space, place, and motion are very familiar to everyone," and have a popular meaning that arises from sense perception, he will give them a technical meaning, and proceeds to describe what he calls "absolute time," whose flow is unchanging, and "absolute space," which remains homogeneous and immovable." Newton then distinguishes between absolute and relative motion. "Absolute motion is the change of position of a body from one absolute place to another; relative motion is change of position from one relative place to another."[20] Thus the motion of a sailor on a ship is the sum of three motions: his motion relative to the ship, the ship's motion relative to the earth, and the earth's motion relative to absolute space.

After the Definitions come the "Axioms, or the Laws of Motion." The first law of motion is of inertia: "Every body perseveres in its state of being at rest or of moving uniformly straight forward, except insofar as it is compelled to change its state by forces impressed." This has been described as "the great factor which in the seventeenth century helped to drive the spirits out of the world and opened the way to a universe that ran like a piece of clockwork."[21] Then comes Newton's second law of motion: "A change in motion is proportional to the motive force impressed and takes place along the straight line in which that force is impressed." Newton thus did not set this down in the form of an equation. Modern notation would express this as $F \propto \Delta(mv)$—or, since change in velocity is acceleration, $F = ma$. The first person to set this down as an equation was Leonhard Euler, almost a century later. Newton's third law of motion: "To any action there is always an opposite and equal reaction."

Why does Newton call these axioms *or* laws? The word "axioms" suggests that what follows is true a priori (by definition or prior to experience), while "laws" suggests that what follows is an empirical fact, something discovered in each actual experience. In fact, Newton's axiom-laws are both: they are true by definition of all possible motions, *and* empirical truths of laboratory measurements done under select conditions. These axiom-laws thus sketch out the meaning of objective knowledge, or what does not change despite different conditions. An object such as a baseball has a certain weight, but its weight is different on the moon and on Mars. How can this be? It's the same ball! What is objective, unchanging, about it is not its weight but its mass, which is independent of what gravity it happens to be in. Similarly, when you throw that baseball you can get it moving at a certain velocity, but you can't get a cannonball moving nearly as much even if you throw it just as hard. Physics has not changed: the relationship among force, mass, and acceleration has remained absolutely the same in this and in all other circumstances. Newton's second law of motion states just this "invariance," as scientists now say. This is why Cohen says that "the Second Law of Motion in its manifold applications lies at the very heart of the Newtonian system of physical thought."[22]

In the *Principia*, Newton has in effect expanded Galileo's thought experiment of an infinite plane without resistances, and produced a complete and abstract world-stage. Only forces, motions, and masses appear on it. There are no human purposes, no final causes. But because of this, certain aspects of motion appear and show themselves on this stage more clearly than in the messy human world. Aristotle had not found it necessary or even possible to imagine such a stage; he saw motion thoroughly woven into the world and unable to be understood apart from it. (In his commentary on the third law of motion, Newton describes a horse pulling a stone tied to a rope, and sets about explaining the forces involved. Imagine how puzzled this would have made Aristotle! He would have asked: Who tied the horse to the stone? Why? What purpose is being actualized here?) Galileo only glimpsed this stage from afar. Newton's stage is spare, but therein lies its beauty and effectiveness. Nature is not to be looked at, as Aristotle had, as a cosmic ecosystem, in which qualitatively different kinds of things act in qualitatively different kinds of ways in qualitatively different domains. It is more like a cosmic billiard table, in which all space is alike, all directions are comparable, all events are motions, and in all changes of motion the same basic kinds of things exert the same basic kinds of forces. In this world, movement involves change in space, not attainment, actualization, or intensification of being. It's a world where all chandeliers, trapezes, and swings are pendulums, all sport and dance instances of $F = ma$, all balls elastic, and all planes go on forever. One can move about this stage anywhere in space and over time—in a car, train, plane, roller coaster, bicycle—and the laws remain the same, "invariant under translation," we would say. If you want to understand what's happening on this world-stage, according to Newton, here's what to do: First, quantify positions, speeds, and masses. Then follow the forces.[23]

This is how $F = ma$ can be both a definition and an empirically discoverable fact. It is a definition insofar as it is part of the warp and woof of the abstract world-stage. It is a fact insofar as, when connected with our world via the right concepts, assumptions, and measurement techniques, it states a quantitative relationship between values found in a laboratory situation. Theories become the vehicles by which to go back and forth between the world-stage and our own, between its ideal values and the real values of our world. To nonscientists, the abstract world might seem strange, something arbitrary and imposed upon nature, a world of fiction—an effective fiction, perhaps, but an invention nonetheless. To scientists, who have been trained to connect this abstract world with our own through concepts, procedures, and measurement practices, the temptation is just the opposite. They can move so confidently back and forth that they can forget how abstract this world-stage is—as if it were not an invented part of the world they live in.

At this final stop, where $F = ma$ appears, therefore, we have traveled quite a distance from the beginning of the human understanding of motion. We see the heavens and the earth as the same place. We do not see an inherent distinction between natural and violent motion, or between natural and unnatural places, or between different kinds of forces. All things follow the same laws, and if we think that something behaves strangely, we assume that we do not yet see how the laws we know apply to it. Motion is not an action but a state. There is one kind of motion, and circular motions are to be explained as the result of combinations of components. Aristotle may be right that motions in the heavens seem different from motions down here, but that's because the resistances in the heavens are different. Nature is a huge space-time determination of forces, accelerations, and motions whose blueprint we can seek through theories. This allows us to view nature as full of quantitative laws to be found through experimentation.

"[Newton] is our Columbus," Voltaire wrote in 1732, "he led us to a new world."[24] But it is a strange world. It is not found in our own like a concealed continent. Nor is it revealed by instruments, the way tiny worlds are seen in microscopes, or distant and gigantic ones in telescopes. Newton's strange new world was found in our world—but it is not our world, either, nor one we could live in. We humans, even the scientists among us, inhabit what philosophers call the "lived world," amid designs, desires, and purposes: we live in an Aristotelian world. The world Newton discovered is an abstract one that appears by changing what we look at and how we look at it. It's a fishbowl-like world, seen from the outside, closed off from our own, whose events disclose to us much about our world. The equation $F = ma$ is the "soul" of that world, as Wilczek wrote, serving to define its structure, and part and parcel of every event that takes place in it.

This is why $F = ma$ is not as straightforward as it appears. When we learn it we are learning more than we think. We are inheriting the entire journey that led up to it.

INTERLUDE

The Book of Nature

> Philosophy is written in this grand book, the universe, which stands continually open to our gaze. But the book cannot be understood unless one first learns to comprehend the language and read the letters in which it is composed. It is written in the language of mathematics, and its characters are triangles, circles, and other geometrical figures without which it is humanly impossible to understand a single word of it; without these, one wanders about in a dark labyrinth.
>
> —Galileo, *The Assayer*

In 1623, Galileo crafted a famous image that is still often cited by scientists. Nature, he wrote, is a book written in "the language of mathematics," and if we cannot understand that language, we are doomed to wander about as if "in a dark labyrinth."

Like other metaphors, this one is both true and untrue; it is insightful but it may be misleading if taken literally. It captures our sense that nature's truths are somehow imposed on us—that they are already imprinted in the world—and underlines the key role played by mathematics in expressing those truths. But Galileo devised the image for a specific purpose. Taken out of its historical context and placed in ours, the image can be dangerously deceptive.

The idea of a book of nature did not begin with Galileo. For centuries it had been an accepted part of religious doctrine that the world contained two fundamental books. Nature, the first book, is full of signs that reveal a deeper meaning when interpreted according to scripture, the second book, which supplies the ultimate meaning or syntax of nature's signs. Understanding involved reading the books together, going back and forth between what one finds in the world and what one reads in scripture. As Peter Harrison has pointed out in his book *The Bible, Protestantism, and the Rise of Natural Science*, reading the Bible was once considered part and parcel of studying nature, and it is therefore wrong to equate serious Bible reading with literalism and antiscientific behavior as we often do today.

During the Renaissance, however, scholars came to appreciate more keenly that the truths of nature were not always easy to discern. Rather, such truths were often cleverly encoded in nature and so required a special training to unlock. Meanwhile, the Protestant Reformation brought about changes in the understanding of texts, emphasizing the truths in them that were exact and self-contained rather than symbolic or allegorical.

Galileo, building on these scientific and religious changes, then appropriated the "two books" image for his own purposes, transforming its meaning.

For in 1623, Galileo was in a jam. His troubles had begun 10 years earlier, when a student of his had discussed his work at the Pisan court, and a participant noted the apparent conflict between scripture and Galileo's scientific claims, especially regarding the motion of the earth. Meanwhile, the authorities were threatening to put *De Revolutionibus*, by his intellectual ally Copernicus, on the official index of forbidden books for similar reasons. Worried for himself and for other scientists, Galileo wrote a letter to the Grand Duchess Christina about the connection between science and scripture. In that letter, he appealed to the traditional image that God reveals himself to humanity in two books, nature and scripture. He suggested that both books express eternal truths and are compatible because they have the same Author—God is saying the same thing in two different ways.

Yet Galileo's novel crafting of the image would prove explosive. Galileo had insisted that the book of nature was not written in ordinary words; its characters were fundamentally different from the words of the scriptures, of Aristotle, and of any textual author. "It is necessary for the Bible," Galileo said, though he might as well have said it of the books of Aristotle, of the Church Fathers, or of any author, "in order to be accommodated to the understanding of every man, to speak many things which appear to differ from the absolute truth so far as the bare meaning of the words is concerned. But Nature, on the other hand, is inexorable and immutable; she never transgresses the laws imposed upon her, or cares a whit whether her abstruse reasons and methods of operation are understandable to men."[1]

Galileo's arguments seem to have convinced Christina, but not the authorities. In 1616, *De Revolutionibus* was put on the Index, followed by Kepler's textbook on Copernican astronomy, *Epitome*, in 1619, and Galileo himself came under attack. Partly in response he wrote *The Assayer*, containing the famous passage that "the grand book of the universe . . . cannot be understood unless one first learns to comprehend the language and to read the alphabet in which it is composed. . . . the language of mathematics." Those versed in mathematics and physics, in other words, can know aspects of God's handiwork that others cannot.

Galileo chose his image carefully, and its roots were deep in Western metaphysics and theology. First, it used the traditional idea that God revealed his power, glory, and truth in the world. Second, it relied on the equally traditional notion that the Bible cannot go against clear demonstrations of logic or the senses. Finally, it appealed to the time-honored metaphor of nature as a book. Galileo was on solid theological ground.

In fact, however, Galileo—perhaps without his being fully aware of it—had stood the old image on its head. The image of the book of nature now implied something almost opposite what it had before—that the signs of nature had their own self-contained meaning. To understand nature one did not need to rely on the Bible as an allegorical aid; studying nature was an independent activity best carried out by a separate, professional class of scholars. If anything, the book of nature now became the primary text—the blueprint written in technical language—and scripture the user's manual, written in popular language.

Galileo thereby used the image to defend not only himself but also all scientists, suggesting that they were as authoritative as the clergy. "The book of nature and those natural philosophers who interpreted it . . . assumed part of the role previously played by the sacraments and the ordained priesthood," writes Harrison.

But the image of the book of nature can haunt us today. One reason is that it implies the existence of an ultimate coherent truth—a complete text or "final theory." While many scientists may believe this, it is ultimately only a belief, and it is far likelier that we will endlessly find more in nature as our concepts and technology continue to evolve. Furthermore, the image suggests that the "text" of the book of nature has a divine origin. The idea that the world was the oeuvre of a superhuman author was the precursor of the idea that it was the engineering project of an intelligent designer. This implication has led some contemporary sociologists of science to succumb to the temptation of characterizing scientists as behaving, and seeking to behave, in a priestlike manner.

The most important lesson to be found in Galileo's image is the need to keep developing and revising the metaphors with which we speak about science.

CHAPTER THREE

"The High Point of the Scientific Revolution": Newton's Law of Universal Gravitation

$$F_g = \frac{Gm_1m_2}{r^2}$$

DESCRIPTION: Gravity exists in all bodies universally, and its strength between two bodies depends on their masses and inversely as the square of the distance between their centers.
DISCOVERER: Isaac Newton
DATE: 1684–87

The high point of the Scientific Revolution was Isaac Newton's discovery of the law of universal gravitation. All objects attract each other with a force directly proportional to the product of their masses and inversely proportional to the square of their separation. By subsuming under a single mathematical law the chief physical phenomena of the observable universe Newton demonstrated that terrestrial physics and celestial physics are one and the same.

—I. Bernard Cohen, *Scientific American*

Just as surely as people know that if you push an object it moves, they also know that, if you drop things like apples, they fall to the ground. No one had to discover this behavior. But Newton's equation—first published not in the form of the familiar equation

$F_g = Gm_1m_2/r^2$ but as a verbal description—*was* a discovery. And it did more than quantify falling behavior, stating the key quantities involved and how they relate. The appearance of this relation—in Newton's *Principia*, the same book in which he published his second law—was the culminating moment of the Scientific Revolution, as Cohen said, for it knit together heavens and earth as part of the same world and obeying the same laws. But the impact of this equation extended yet further. It helped enshrine Newton as a symbol, not only of scientist, explorer, and genius, but—strangely enough, given that the Aristotelian scientific picture was being ushered off the horizon—also of humanity's quest for actualization and perfection: *This* is what we can accomplish when our minds are fully engaged. Indeed, Newton's discovery of universal gravitation seemed a close encounter with divinity: *This* is as close to God as we humans can ever hope to get. It is thus not a coincidence that Newton's discovery is firmly connected with a story involving an apple, recollecting that other famous apple story—the biblical story of the Garden of Eden, and the first fruit of the Tree of Knowledge to be grasped by humans.

"The Most Difficult Question in Physics"

In Aristotle's cosmic ecosystem, falling was a special behavior that only certain kinds of things did, and only in certain places in the universe. Falling was one among many different kinds of motion and change, and had nothing to do with the tides, nor with the circular motions of the planets and other heavenly objects. It was a natural motion by which a thing made of some proportion of earth returned to its natural place via its own internal power. The causes of falling therefore included the composition of the object, its natural place in the earth, and the object's tendency to return to that place. For a long time, under Aristotle's influence, the downward falling of objects to the earth was viewed as but one of several different types of "attractions" and motions in the universe. So was his view that the quickness of fall depends on the heaviness of the object—which is, after all, confirmed by our everyday experience. As the character Rosencrantz, holding up a ball and feather at one point in the movie of Tom Stoppard's play *Rosencrantz and Guildenstern Are Dead*, says, "You would think this would fall faster than this [drops them, ball hits the ground first]. And you would be absolutely right."

But some ancient authors broke with Aristotle in proposing the existence of various kinds of connections between phenomena on earth and in the heavens, the most conspicuous being that between the moon and the tides. Aristotle had struggled to produce a mechanical explanation for tidal motions—involving the wind—but others thought the connection somehow more direct. The Greek scholar Posidonius (ca. 135–51 BC), along with several other ancient authors, produced the un-Aristotelian notion of forces permeating the cosmos that were not based in any one substance (substances being the only things that truly existed for Aristotle), but which linked substances together. These cosmic forces were called "sympathies," after the Greek for "feeling together."[1]

In the ancient and medieval world, the exploration of physical influences among heavenly bodies, and between the heavenly bodies and objects on earth, was generally called "astrology." But we must not confuse this with the current socially acceptable form of bigotry that seems to entitle the human beings who believe in it to prejudge the character of others based solely on their dates of birth. Ancient and medieval astrology indeed had its share of charlatans who did that sort of thing. But astrology also had a serious side, springing from the quite reasonable assumption that physical influences existed in the universe that linked some things to other faraway things, and the scholarly conviction that it was possible to investigate and describe these influences. As science historian David C. Lindberg says, "Almost any ancient philosopher would have considered it extraordinarily foolish to deny the existence of such connections."[2] The work of astrologers, at the beginning, had an enormous positive result in developing notions of long-range forces.[3]

Yet the problem of explaining these connections, including why bodies fell, remained puzzling. Was the force something external or internal to the falling object, or something else? In 1504, Nicoletto Vernias, writing on free fall, declared, "This question is the most difficult of all questions in physics."[4]

The question was transformed in 1543, when Nicolaus Copernicus (1473–1543) published *On the Revolutions of the Heavenly Spheres*, a book proposing that the sun, not the earth, was the center of the solar system. This book—according to legend the author received the first published copy on his deathbed—assumed that gravity was a volition implanted by God into things. Yet it profoundly influenced those who investigated cosmic forces, inasmuch as it implied that the gravity or heaviness of bodies on earth was not cosmically unique but presumably experienced on the other bodies orbiting the sun—and perhaps by the moon and even the sun.[5] Every body had its own gravity.

Another milestone in thinking about cosmic forces was *De Magnete* (1600), William Gilbert's treatise on magnetism. Magnetism clearly sprang from the mutual interaction between the earth and various substances, and Gilbert noted that its strength varied with distance. He also suspected that magnetic force was active even when the bodies it affected were at rest. Gilbert ridiculed the notion that bodies at rest were unaffected by this force as like thinking that houses are governed by walls, roof, and floor rather than the families inhabiting them.

The work of Johannes Kepler (1571–1630), who sought a mathematical description of the cosmic force binding the sun and planets, built on that of Copernicus and Gilbert. Kepler's early studies in theology were unexpectedly interrupted when he landed a job as a mathematician, and he developed an ambivalent relationship with astrology. Like the astrologers, he was passionately devoted to the idea that harmonies pervaded the universe, anchored in an overarching harmony established by God. Yet Kepler scorned the methods of astrologers, for they were firmly committed to everyday language

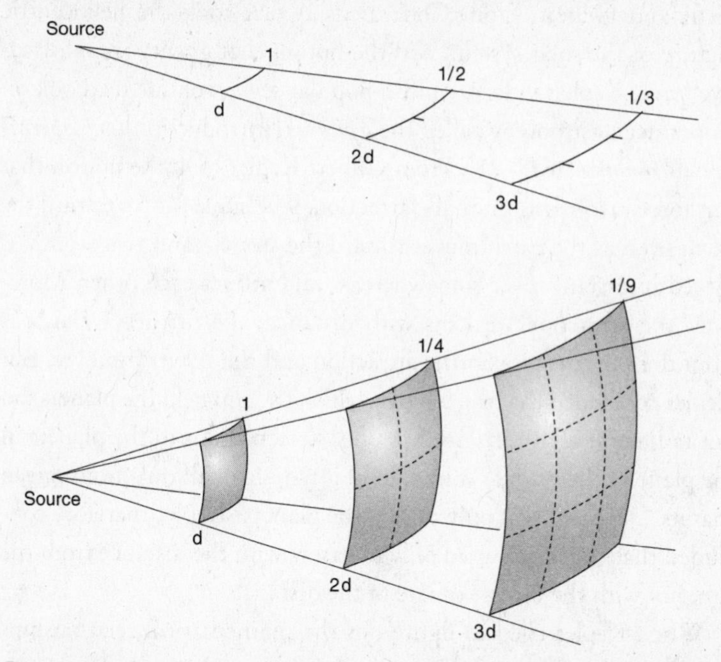

The strength of a force that extends out in a plane will weaken directly as the distance from the source, while that which radiates in all directions will weaken with the square of the distance: the inverse square law.

in their studies, and knew nothing of the precise language of mathematics used by professional astronomers. Without mathematics, Kepler thought, astrologers could not detect the cosmic harmonies, and would be ignorant of the structure of the world.

Kepler was among the first, for instance, to realize that the intensity of light varies according to an inverse square law. An inverse square law states that some property weakens with the square of the distance. In the case of the intensity of light, which radiates out in all directions from a source, it is simple geometry. If you double the distance from a source, for instance, the area over which the same light must be distributed increases (and its intensity weakens) by four; if you triple the distance, the area increases by nine.

Kepler's astronomical work borrowed elements from both Copernicus and Gilbert. From Copernicus Kepler took the heliocentric picture of the solar system, and the notion that gravity is an attractive force; Kepler, in fact, wrote a popular seven-volume textbook on Copernican astronomy called the *Epitome* [Introduction] *to Copernican Astronomy* (1618–21). From Gilbert Kepler took the notion that this force involves a "mutual" attraction. The stone moves toward the earth even as the earth moves toward the stone—and two stones, if placed in distant space somewhere, would attract each other. Moreover, the attraction weakens with distance; the further a planet is from the sun, the weaker the attraction and the slower it moves. But Kepler concluded that the force by which the sun held the planets did not radiate in all directions, but only stretched out to the planets in the plane of their orbit. Why should it radiate in all directions, given that its "purpose" was only to grip the planets? Kepler therefore concluded that the force varied only inversely with the distance from the sun, not with the inverse square of the distance.

When Kepler tried to figure out the mathematical relationships governing the planetary motions, though, he encountered a puzzle. According to Copernicus, the planets revolve about the sun, and in circular orbits, for all celestial motions had been considered circular since the time of Aristotle. But before the use of telescopes in astronomy began in 1609, the best data of the day had been taken by the Danish astronomer Tycho Brahe (1546–1601), whom Kepler knew and trusted—and Kepler found that these data could not *quite* be fitted to a circular orbit model. The discrepancy was tiny, almost insignificant, a mere 8 minutes of arc, or just barely more than the naked eye was able to discriminate. Kepler spent six years trying to incorporate those 8 minutes of arc into the Copernican system. He could not.

Others might have written off the discrepancy as due to observational error or to some unknown factor. Yet Kepler trusted *both* Copernicus's heliocentric model *and* Brahe's data. Because he did, he was led to consider a radically new idea: that the planets move, not circularly, but in elliptical orbits with the sun at one focus. Furthermore, he concluded that, regardless of whether the planets move quickly when near the sun, or more slowly when more distant, an imaginary line stretching from the sun to a planet sweeps out equal areas in equal times. These conclusions were the first two of Kepler's famous three laws, and the third was another mathematical relationship: that the squares of the times of revolution of any two planets are proportional to the cubes of their distances from the sun.[6]

Kepler found these laws beautiful and harmonious. He also claimed that this beauty and harmony was what had caused God to use these laws to construct the universe in the first place. "This notion of causality," notes philosopher E. A. Burtt, "is substantially the Aristotelian formal cause reinterpreted in terms of exact mathematics."[7] And Kepler saw the force binding sun and planets as a secular version of an animate force. "If for the word 'soul' you substitute the word 'force,' " he wrote, "you have the very same principle on which [my] Celestial Physics [is based]. . . . For once I believed that the cause which moved the planets was precisely a soul. . . . But when I pondered that this moving cause grows weaker with distance . . . I concluded that this force is something corporeal."[8] This almost seamless transition in Kepler's thinking between the sun gripping the planets like a spirit and gripping them with a corporeal force is a classic illustration of what French philosopher Auguste Comte called the transition between theological and metaphysical thinking.

But what kind of thing was this corporeal force? This question would be debated for most of the rest of the seventeenth century. Some agreed with Kepler that it was a corporeal force. Others, such as Descartes, thought that it was purely mechanical and a product of tiny motions, called vortices, in a fluidlike substance called the ether in which the solar system was submerged.[9] Galileo, taking what Comte would have called a step into scientific thinking, preferred to stop discussing the nature of gravity altogether, and focus on measuring its quantitative effects. Just give us the numbers, please.

In 1645, French astronomer Ismael Boulliau (1605–1694) unwittingly stumbled across, and rejected, the right formula for the strength of this force. Boulliau is a fascinating figure in the history of science, known for his accurate astronomical tables and quirky intellectual commitments. He was one of the first astronomers to accept Kepler's idea that the planets move in elliptical orbits—but also one of the last astronomers to take astrology seriously, which led him to attack Kepler and his use of mathematics. On astrological grounds, Boulliau vehemently rejected Kepler's conclusion that planetary motion was governed by an impersonal force from the sun whose strength weakened with distance. If there were such a force, Boulliau proclaimed, laughing at Kepler's ridiculous idea, it would have to spread out in all directions, like light, meaning that it would weaken as the *square* of the distance. But this is absurd! Boulliau could not believe that God would act in such a way.[10]

Several other scientists, however, realized that the force between planets and sun might indeed radiate in all directions, meaning that something like an "inverse square" relationship was not absurd, and indeed probably involved in whatever force operated between sun and planets—but they thought that this relationship was the outcome of a tug of war between a center-fleeing, or centrifugal, force and a center-seeking force, with the inverse square behavior as the outcome. These scientists also suspected that Kepler's laws could be derived from an inverse square relationship.

One was Robert Hooke (1605–1703), the curator of experiments at the Royal Society in London. In 1674, Hooke proposed that the earth and all other celestial bodies possess "an attraction or gravitating power towards their own Centers," which attracts not only parts of that body but all other bodies "within the sphere of their activity," with the strength of the force depending on the distance between the bodies.[11] Yet Hooke did not have the mathematical ability to use this surmise to calculate planetary motions. In 1679, still seeking an answer, he wrote a letter to the ablest mathematician around, Isaac Newton. What, Hooke asked Newton, do you think of my ideas about "an attractive motion towards the central body"?[12] And in January 1680, after exchanging letters with Newton, Hooke mentioned his idea that the attractive force varied according to an inverse square law. If this were so, he asked Newton, would the planetary paths work out?

In 1680, as it happened, several events sparked interest in the motions of celestial objects, and curiosity about their behavior. One was the appearance of a large and dramatic comet in the heavens, which was examined with interest by British astronomer Edmond Halley (1656–1742). That comet was followed by another comet in 1682—this one is now known as "Halley's comet"—and yet another in 1684. Until this time, comets were assumed to be alien, randomly appearing objects in the solar systems, not governed by its laws. This opinion would soon change.

In January 1684, in a London coffee house, Halley, Hooke, and the scientist and architect Sir Christopher Wren (1632–1723) discussed the nature of planetary paths, and whether they could be accounted for by an inverse square law. Halley said he had tried and failed to calculate the paths based on such a law. Hooke boastfully said that he had done it but refused to produce a demonstration. Wren, both skeptical and impatient, challenged them to produce a demonstration within 2 months, saying that he would reward the one who did so with a book worth 40 shillings. The 2-month period expired, but the next time Halley wound up in Cambridge, that August, he broached the subject with Newton. That visit was the single most transformative event in Newton's life. And it begat one of the most important events in Western science and culture—the birth of the *Principia*—in which the law of universal gravitation was a by-product.[13]

"One of the Most Far-Reaching Generalizations of the Human Mind"

Laws are like sausages, runs the old saw: the less you know how they are made, the more you respect the product. This remark is more clever than true. For what do you expect? If you truly understand human creativity, you have no difficulty with knowing how tasty sausages and just regulations are made. But the remark does highlight a curious conundrum of creativity: that something momentous can arise from base origins. Few triumphs of the human mind illustrate this conundrum as sharply as Newton's path to the law of universal gravitation. That path was marked by raging ambition, empty posturing, obsessive secrecy, seething jealousy, and transparent lies, but the product was breathtakingly brilliant. It was, Richard Feynman once said, "one of the most far-reaching generalizations of the human mind."[14]

Newton's path to universal gravitation evolved during the same time period as his path to force discussed in the previous chapter. It began when he was a student at Trinity College, Cambridge, and jotted down numerous remarks about gravity in his notebooks. In some, he treats gravity as if it were an impetus-like ability internal to things that caused their motion; in others, dealing with celestial motions, he considers Descartes' explanation that gravity arose from the pressures of particles created by vortices. For a long time, he accepted the notion of a centrifugal force, one that pushed away from a body, as a swinging stone tugs on the end of the tethered string to which it is attached.

Then, about 1680, Newton's thinking about gravitation was profoundly altered by two key events, one philosophical and the other mathematical. The philosophical event was conversion away from an impetus-like idea of force as something that impelled a body to move from *within*, to the view that motion is caused by a force that acts on a body from *without*. This was accompanied by a dawning realization of the distinction (noticed before him, with varying degrees of clarity, by Robert Boyle, Galileo, and Kepler) between weight and mass, which is necessitated by the idea of forces that vary with distance. Weight varies with the distance from the earth's surface; a body has a different weight at different altitudes. But the mass of a body, which is a key element of how the body moves, stays the same.

The other key event to profoundly alter Newton's thinking about gravitation was the correspondence with Newton's nemesis Hooke that began in the fall of 1679.

Newton loathed Hooke. In 1673, Hooke had told his Royal Society colleagues—mistakenly but pompously—that Newton's recent, pathbreaking work on light was wrong, making Newton so annoyed that he threatened to give up science altogether. The correspondence that Hooke initiated in the fall of 1679 and continued for 2 months began equally inauspiciously. Newton made an embarrassing error in his first reply, and again Hooke bruited about Newton's blunder to his Royal Society cohorts. But Newton was challenged by Hooke's question about the inverse square law and planetary motions. He was also intrigued by Hooke's remark that the planets travel in curved paths, not because of the combined action of centrifugal and centripetal forces acting on them, but because of the combined action of a centripetal force and the bodies' own inertia.

This latter observation "set Newton on the right track," though Newton would spend the rest of his life denying Hooke's contribution.[15] In the early 1680s, Newton did not have the law of universal gravitation yet; for one thing, he still treated comets as aliens to the solar system. But he used Hooke's method of analyzing curved motion by decomposing it into a straight-line centripetal force and straight-line inertial motion to great effect. It opened the door to think of everything—falling bodies, planets—as governed by one center-seeking force. Newton also employed Hooke's method, plus the inverse square law, to establish the fundamental connection of Kepler's laws of motion. A body, attracted by another body by an inverse square force, travels around it in an elliptical path with the central body at one focus, and a line drawn between the central and orbiting bodies sweeps out equal areas in equal times.[16]

Then Halley dropped by to visit Newton in Cambridge in August 1684. A contemporary described the visit:

> After they had been some time together, Dr [Halley] asked him what he thought the curve would be that would be described

> by the planets supposing the force of attraction towards the Sun to be reciprocal to the square of their distance from it. Sir Isaac replied immediately that it would be an ellipsis. The doctor, struck with joy & amazement, asked him how he knew it. Why, saith he, I have calculated it. Whereupon Dr. Halley asked him for his calculation without any further delay. Sir Isaac looked among his papers but could not find it, but he promised him to renew it and then to send it to him.[17]

Was this Newton's paranoia and secretiveness, or had he really misplaced the calculation? We can't say. In any case, Newton set out to rework the calculation for Halley, and by early December this transmuted into the first draft of a short, nine-page work entitled *De Motu* (*Concerning Motion*). In it, Newton treated the sun as fixed and immobile: as a body that attracted everything else in the solar system but remained unaffected by the planets swirling about it. This work brought Newton to the threshold of universal gravitation, but it lacked a key idea. According to Newton's third law of motion—for any action there is an equal and opposite reaction—if the sun tugged on a planet it meant that the planet also tugged back on the sun, affecting its motion. This seems to have occurred to Newton only after completing the first draft of *De Motu*.

Newton therefore set about revising the work, which he finished by the end of December 1684. This is the first document to embody the key insight of universal gravitation—that all bodies act on each other—using the phrase "eorum omnium actiones in se invicem," or "the actions of all these on each other." If the sun had one planet orbiting about it, for instance, the two bodies would revolve about a common center of gravity. But the solar system contains many planets, each of which tugs on the sun and on one another. No planet therefore moves in a perfect ellipse, nor ever follows the same path twice. Indeed, Newton wrote, to calculate the complex net result of all the tugs "exceeds, unless I am mistaken, the reach of the entire human intellect."[18]

Newton had not only achieved a deeper insight into the solar system but had also transformed scientific procedure. He had transformed Galileo's thought experiment of an infinite plane without resistances into a complete world-stage, on which masses appear and do nothing but move under the influence of forces. Scientists create models on this world-stage—such as Kepler's laws of motion—and compare these models to observations of the real world. But these models are only approximations, and have to be constantly refined. Newton's early work had been motivated by Kepler's laws, which he had assumed to be accurate descriptions—and these had led him to conclude that Kepler's laws were wrong, and to predict deviations from them.[19]

Newton gave *De Motu* to Halley in December 1684. Halley asked Newton if he could publish it, but Newton refused. Instead, he set about expanding *De Motu*, weaving together his new insights into the structure of the solar system with other insights, including Hooke's idea of analyzing circular motion into two components.

The result, which appeared after 18 months of labor in 1686, was the *Principia*, the single most influential piece of writing in science. Near the beginning of Book I, Newton skillfully uses Hooke's method of analyzing curved motions by breaking them down into centripetal forces plus inertia, to derive Kepler's laws, among others. In a part of Book II, Newton demonstrates that Descartes' vortices could not explain the motions of planets, and promises an adequate explanation. In Book III, the "System of the World," Newton follows through on that promise. He carries out the "moon test," measuring the force tugging objects on the earth's surface, and shows that it has the same strength as the force with which the earth tugs the moon; furthermore, this force has the same strength as that between the sun and the planets, and the other planets and their satellites. Until now, Newton announces dramatically a few paragraphs later, we have called these all "centripetal" forces—but now that we are sure it is the same force, we can call it by one name: gravity. Gravity "exists in all bodies universally," and its strength between two bodies depends on their masses and "will be inversely as the square of the distance between the centers." As we write it now, $F_g = Gm_1m_2/r^2$.

Hooke later claimed priority for the discovery of this law, and we can see why. But we can also see why Newton (and many historians) rejected this claim. Newton clearly profited from Hooke's work, but when Newton famously said that he saw farther than others because he stood on the shoulders of giants, the statement owed its truth to its irony. Newton was alluding sarcastically to Hooke's diminutive stature; that is, the boost from him was more like that of a footstool than a tower. Hooke had suggested the inverse square law chiefly with respect to one body, or at most with respect to celestial bodies, while Newton made its universality explicit. Hooke's most important boost to Newton had been in showing Newton how to analyze curved orbital motions. But the priority issue is further obscured, both factually and morally, by Newton's mendacious practice in memoirs and conversations of backdating key events in his work on universal gravitation—including the moon test—to clinch priority over Hooke. Still, what makes Newton stand out as its discoverer—above Boulliau, Hooke, and others—is his clear statement that gravity is not just a force by which certain bodies grip or are gripped by certain other bodies—not bodies falling and the bodies to which they fall, nor heavenly bodies to one another—but all bodies to all other bodies.

There were some odd features of Newton's account. Why, for instance, was the mass of a body involved in the gravitational force the same as the mass in the push-pull force described by $F = ma$? It didn't have to be. Was this an accident? If so, it was an awfully strange accident. The answer to this puzzle would play a role in the development of general relativity over 200 years later. But in Newton's time, one would have to think carefully to see it as a puzzle, so daring and dazzling was the sweep of Newton's vision.

It was a deeply democratic vision. Gravity is a universal force, and it does not matter what a body looks like, nor where it lives in the universe, but solely how much mass it has. Galileo had universalized things, and achieved insights, by turning all chandeliers into pendulums. Newton now universalized even more ambitiously by turning all bodies into attractors. Gravitation is all bodies, all the time, everywhere.

The Law That Explained Law

Newton's equation of universal gravitation was hailed as the capstone of one of the most profound transformations of Western science. It led to Newton becoming the "gold standard" against which scholars in other sciences compared the superstars in their fields. James Clerk Maxwell, for example, hailed Ampère as the Newton of electricity, while Alfred R. Wallace, Thomas Huxley, and others called Darwin the Newton of biology.

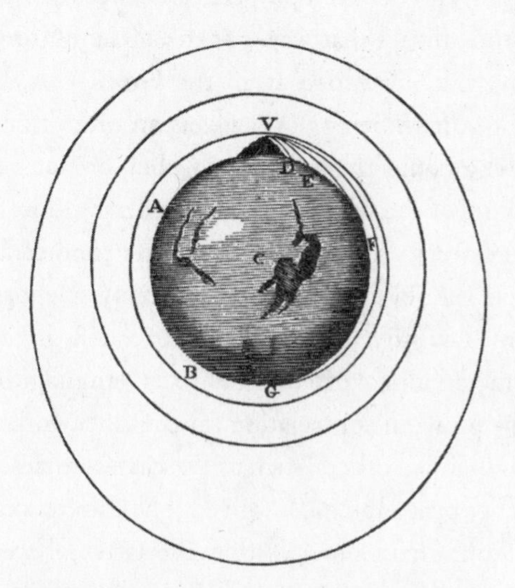

Diagram of Newton's cannonball thought experiment illustrating the idea of an orbit. What would happen if one shot a cannonball horizontally from a peak that poked above the atmosphere? The more forcefully the cannonball is shot, the farther around the earth it will travel. With enough force, it returns to the peak, and follows the same path over and over again.

Moreover, Newton's law of gravitation was often cited as the kind of law that a mature science required. François Magendie, in his classic textbook *Elementary Sketch of Physiology* (1817), lamented the absence from his field of "an intellect of the first order to come and discover the laws of the vital force in the same way Newton made known the laws of attraction."

But the influence of Newton's equation extended well beyond science—to education, philosophy, theology, and other areas of human culture. It also helped to change the very notion of "law" itself.

In modern times, the concept of a scientific law has a specific meaning; it is something descriptive of nature and its behavior. For example, in his book *The Software of the Universe: An Introduction to the History and Philosophy of Laws of Nature*, the philosopher Mauro Dorato from the University of Rome 3 calls a scientific law "a mathematical relationship between properties of physical systems."

But it was not always that way. For the ancient Greeks, laws were not descriptive but *normative*, from the Greek *nomos*, the custom or behavior of human beings. A law was an order that a ruler gave to subjects, who could then choose to obey or not obey. (For the nonhuman parts of the world, what we understand in terms of laws was then expressed in the idea of the thing's characteristic nature.) Even as late as the seventeenth century, many scientists refused to apply the term "law" to regularities in nature, insisting that it was no more than a metaphorical extension of social language to the natural world. But the growing appreciation for the clockworklike structure of the cosmos inclined others, such as Descartes, to describe creation as a juridical act by a supreme lawgiver. The difference between the human and nonhuman order is that the latter obey God unconsciously, while the former obey (or disobey) consciously.

Newton viewed the world in this manner. He saw himself as describing a universal principle that pervades the *entire* universe and affects *everything* in it, something whose influence is direct, immediate, and authoritative. The very universality of this principle, and the care with which Newton states that gravitation is not a property in matter, was part and parcel of his view that he was describing the actions of a supreme lawgiver.[20] Newton's matter is lifeless; it moves only when touched by a force. This got "out of God's way," to guarantee that the Creator had a free hand.[21] Newton's mechanical view of the universe, as full of objects passively responding to forces from without, thus was not only consistent with a supreme lawgiver, but required it. How could there be laws and no lawgiver? As he wrote, "This most Elegant System of the Planets and Comets could not be produced but by and under the Contrivance and Domination of an Intelligent and Powerful Being." (That Sir Isaac Newton could have thought this regarding the origin of the solar system, which is now easily accounted for by the action of simple principles over time, makes us marvel at the outrageous hubris of those much smaller intellects today who are so confident that their inability to explain something's origin means that this thing must have been the act of a god.)

Newton's equation of gravitation gave an enormous boost to the inclination to view laws descriptively rather than normatively. The influence was reversed: natural language was now extended to the social world.

One of Newton's assistants, the Royal Society member John Theophilus Desaguliers, composed a poem entitled "The Newtonian System of the World, the Best Model of Government." Desaguliers found in the Newtonian system, consisting of "the most regular Attraction of universal Gravity, (or attraction) whose Power is diffus'd from the Sun to the very Centers of all the Planets and Comets" to be a "lively image of our System" of government (the British), namely, "The limited Monarchy, whereby our Liberties, Rights and Privileges are so well secured." Thanks to this, he concluded, "the Happiness that we enjoy under His present MAJESTY's Government" is a sign "that A-T-T-R-A-C-T-I-O-N is now as universal in the Political, as the Philosophical World."[22]

But political theorists also began to use Newtonian language—so much so that it sharply influenced the modern conception of democracy, as Cohen detailed in his 1995 book *Science and the Founding Fathers: Science in the Political Thought of Thomas Jefferson, Benjamin Franklin, John Adams, and James Madison*. All of the U.S. founding fathers read Newton, Cohen pointed out. Jefferson, whom Cohen describes as "surely the only president of the United States who ever read Newton's *Principia*," had several copies of the *Principia* in his library and Newton's portrait on the wall; Franklin was so deeply impressed by Newton as a young man that he tried to meet him in London; Adams once cited Newton's laws of motion in a political debate; and Madison wrote an essay comparing nature and human affairs.

Even the birth of socialism is tied up with Newton's law. For the political thinker Henri de Saint-Simon (1760–1825), who was one of the founders of socialism, Newton's law was not only the purest example of scientific thinking but also provided the model for creating a science of human social life, based on universal fraternity and collective organization. Saint-Simon once had a vision in which God disclosed to him that Newton sat at his right hand and decreed that the world should be governed by a committee called the Council of Newton. Its primary task, besides improving humanity—Saint-Simon is quoting God now—was to discover "a new law of gravitation applicable to social bodies." Newton's equation was not just a key fact, but *the* key fact, unifying science and provoking the search for a law of social order that would work not merely between individuals and groups but also nations. Saint-Simon even criticized Newton for failing to turn gravity into an all-encompassing philosophical system.[23] The sooner humanity found this law and reorganized society accordingly, the sooner it would be liberated.

To be sure, Saint-Simon was a flamboyant character, and the kind of megalomaniac aristocrat—idealist, bad writer, idiot, and eccentric—with which early nineteenth-century socialism was amply stocked. But he was not alone. Other political thinkers, including Pierre Cabanis (1757–1808), Charles Fourier (1772–1837), and Giovanni Morelli (1816–1891), tried to apply the notion of gravitational attraction to human life in holding that free, subjective, conscious individuals were nonetheless compelled by universal, deterministic scientific law—a notion that also influenced Karl Marx (1818–1883).

Newton's equation of universal gravitation did more than quantify the attraction between objects, be they pebbles, spacecraft, or planets. Among other things, it inspired scholars in other fields—even political theory—to seek descriptive, mathematical, and universal laws. If the Pythagorean theorem was a proof that exhibited Proof, Newton's equation of universal gravitation was a law that exhibited Law. In so doing, it not only altered our understanding of nature, but also our conception of science and human life.

The equation remains a symbol of the achievement of knowledge and rationality. In George Orwell's novel *1984*, the final sign that protagonist Winston Smith (after accepting that 2 + 2 = 5) had fully capitulated to the thought police—had been thoroughly broken and ceased to think—is that he denies the law of gravity.

INTERLUDE

That Apple

Then ye who now on heavenly nectar fare,
Come celebrate with me in song the name
Of Newton, to the Muses dear; for he
Unlocked the hidden treasuries of Truth:
So richly through his mind had Phoebus cast
The radiance of his own divinity.
Nearer the gods no mortal may approach.

—Edmond Halley, *Ode to Newton*

What of the apple?

The story that Newton discovered universal gravitation after seeing an apple fall is one of the oldest and most familiar legends of science.[1] The incident is said to have taken place some time late in 1665 or 1666 at his mother's orchard in Woolsthorpe, Lincolnshire, where Newton had retreated from studies at Cambridge to escape the plague. The story has long been dismissed as fiction, for several reasons. First, it seems just too theatrical to be true. Second, a cranky but influential early biographer named David Brewster doubted the story. Third, and most importantly, the story is just not how great revolutions happen. The causal force implied by the story—that seeing an apple fall created the law of universal gravitation in Newton's mind, without much further analysis and reflection—has to be false. As Newton's biographer Richard Westfall observes, "The story vulgarizes universal gravitation by treating it as a bright idea."[2]

Yet biographers have found abundant evidence that the ultimate source is Sir Isaac himself, who told the story to several different people—including his niece (who passed it on to Voltaire) and to his friend William Stukeley (1687–1765). Here's the version from Stukeley's memoirs:

> The weather being warm [Kensington, England, April 15, 1726], we went into the garden and drank tea, under shade of some apple-trees, only he and myself. Amidst other discourses, he told me, he was just in the same situation, as when formerly, the notion of gravitation came into his mind. It was occasion'd by the fall of an apple, as he sat in contemplative mood. Why should that apple always descend perpendicularly to the ground, thought he to himself. Why should it not go sideways or upwards, but constantly to the earth's centre?[3]

But we should still be skeptical. Why should this notoriously shy, secretive, and possessive person suddenly become garrulous, expansive, and giving about the origins of his greatest discovery? It doesn't sound like the genuine Newton. Many writers and historians, in fact, suspect that it was not—that he was being devious, in an attempt to attack Hooke. For Hooke had claimed to be the first to come up with the inverse square law for gravitation, and had even once written a letter to Newton seeking Newton's approval of the claim that he, Hooke, had come up with the law first. In telling this story, Newton was predating his discovery of gravitation to the 1660s, and thus removing the ground from Hooke's claim. Such a deception sounds more like the genuine Newton, even if it is not the truthful Newton.

For the truthful Newton, the best we have is the following response he made when once asked how he made discoveries such as the law of gravitation: "By always thinking [about] them," Newton said. "I keep the subject constantly before me and wait till the first dawnings open little by little into the full light."[4] This rings far truer, and more closely resembles how other great discoveries are made, than the apple story. The dawnings involve more than seeing clearly, but conceptual shifts, transforming what he had inherited, and forging new things and new concepts. "The momentous discovery of universal gravitation, which became the paradigm of successful science, was not the result of an isolated flash of genius," wrote I. Bernard Cohen, but a lengthy process involving the "transformation of existing ideas." Cohen added, "The discovery of universal gravity brings out what I believe is a fundamental characteristic of all great breakthroughs in science from the simplest innovations to the most dramatic revolutions: the creation of something new by the transformation of existing notions."[5]

An apple may well have played a role in Newton's thinking of gravitation. But if it did, it served a similar purpose to Socrates' pointing to the diagonal did to Meno's slave—it helped him recast what he already knew of the problem in a new light, helping to transform that thinking in the process.

CHAPTER FOUR

"The Gold Standard for Mathematical Beauty": Euler's Equation

$$e^{i\pi} + 1 = 0$$

DESCRIPTION: The base of natural logarithms (an irrational number) raised to the power of pi (another irrational number) multiplied by the square root of negative one (an imaginary number) plus one is an integer: zero.
DISCOVERER: Leonhard Euler
DATE: 1740s

> Like a Shakespearean sonnet that captures the very essence of love, or a painting that brings out the beauty of the human form that is far more than just skin deep, Euler's equation reaches down into the very depths of existence.
>
> —Keith Devlin

When the 14-year-old Richard Feynman first encountered $e^{i\pi} + 1 = 0$, the future Nobel laureate in physics wrote in big, bold letters in his diary that it was "the most remarkable formula in math." Stanford University mathematics professor Keith Devlin writes that "this equation is the mathematical analogue of Leonardo da Vinci's Mona Lisa painting or Michelangelo's statue of David." Paul J. Nahin, a professor of electrical engineering, writes in his book, *Dr. Euler's Fabulous Formula*, that the expression sets "the gold standard for mathematical beauty." One of my correspondents said it was "mind-blowing"; another called it "God's equation."

This expression, discovered by the eighteenth-century Swiss mathematician Leonhard Euler, has become an icon—an object with special properties above and beyond the truths that it represents—for many people, even those with only a little mathematical training. Like other icons, it can become an object not only of fascination but also of obsession.

Consider first that it is surely one of the few mathematical expressions to serve as a piece of evidence in a criminal trial. In August 2003, an ecoterrorist assault on a series of car dealerships in the Los Angeles area resulted in $2.3 million worth of damage; a building was burned and over 100 SUVs were destroyed or defaced. The vandalism included graffiti consisting of slogans such as "GAS GUZZLER" and "KILLER"; and, on one Mitsubishi Montero, the formula $e^{i\pi} + 1 = 0$. Using this as a clue and later as evidence, the FBI arrested William Cottrell, a graduate student in theoretical physics at the California Institute of Technology, on eight counts of arson and conspiracy to commit arson. At the trial that resulted in his conviction, in November 2004, Cottrell admitted having written that equation on the Montero. "I think I've known Euler's theorem since I was five," Cottrell said during the trial. "Everyone should know Euler's theorem."[1]

Another equation-turned-icon, and one certainly much better known than Euler's, is $E = mc^2$. This equation is a familiar part of popular culture, and has even been turned into a monument: during the 2006 World Cup, the six large outdoor sculptures that were erected in Berlin to illustrate Germany's status as the "land of ideas" included a car, a pair of football boots, and a gigantic representation of $E = mc^2$.

But how is it possible for an equation to become an icon, anyway? After all, an equation is merely one step in the ongoing process of scientific inquiry. Euler's expression was but one implication of his exploration of functions, while $E = mc^2$ was an afterthought of Einstein's development of special relativity. Aren't equations just tools of science, of less intrinsic value and interest than the tasks they were developed to help us with? How do some of them acquire an inherent value or significance beyond the process of inquiry to which they belong? Tools surely can become icons, the way, for instance, a hammer and sickle became symbolic of the Soviet state—but a mathematical and technical object like an equation? What makes such an abstract thing able to stand literally alongside a pair of boots or a car?

The story of Euler's formula helps to answer these questions.

The Neighborhoods of Mathematics

Leonhard Euler (1707–1783) was the most prolific mathematician of all time; his collected works, when finished, will run to some seventy-five volumes. He calculated effortlessly, "just as men breathe, as eagles sustain themselves in the air."[2] It helped that he had a prodigious memory that spanned the swath of human knowledge, able to retain extensive mathematical tables and the entire text of Virgil's *Aeneid*. It also helped that he had an eye for spotting deep connections between what seemed to be vastly different areas of mathematics, synthesizing them and making the result seem as obvious as $2 + 2 = 4$. His equations about fundamental matters are of such simplicity and elegance that, one commentator remarked, their "form pleases the eyes as much as the spirit."[3] His famous formula $e^{i\pi} + 1 = 0$ was the most simple, elegant, and pleasing of all.

Euler was born in Basel, Switzerland. His father, a Protestant minister, awakened his earliest interest in mathematics by instructing him in the basics. Euler continued to receive private math tutoring in high school, because the subject was not taught there. At age fourteen he entered the University of Basel and studied a wide range of topics, from theology to languages to medicine, but remained fascinated by math. Saturday afternoons he was privately coached by the renowned mathematician Johann Bernoulli, and became friends with the latter's sons Nicolaus and Daniel. After Euler received his degree, in 1723, he complied with his father's wishes and tried to become a theologian, but soon turned back to mathematics.

Math was not an easy career. Universities then were bastions of scholarship in the humanities, with few places for mathematicians or scientists. The rare available jobs for mathematicians were at a handful of royal academies.

Leonhard Euler (1707–1783)

Fortunately for Euler, Peter the Great of Russia and his second wife, Catherine I, one of history's great "Renaissance couples," were in the process of founding the Russian Academy of Sciences in St. Petersburg, and plucking for it leading scientists from all over Europe. Two early recruits were Nicolaus and Daniel Bernoulli, who in turn secured an invitation for their friend Euler. Both Peter the Great and Catherine died before Euler arrived, in 1727, and their successors were less enthusiastic about the academy; still, Euler was well cared for and supported. He was surrounded by first-rate scientists and was soon the academy's chief mathematician. He was so productive that the editors of the academy's journal stacked his manuscripts in piles, grabbing a few from the top when they had space. His 14 years at the academy were accompanied by some hardships, the worst of which was the loss of his right eye, probably through eyestrain due to overwork. But during these years he was free to calculate furiously, and reshaped the foundations of mathematics in the process.

Mathematics often grows in an indirect way, the way that many cities do. Certain scattered settlements spring up first, with little interaction among one another. These settlements eventually cluster around one another, becoming neighborhoods, but because they form almost at random they are poorly adapted and little commerce takes place. A visionary leader emerges who understands each neighborhood, and by renaming some streets and building others between key centers allows them to grow into a greater structure that is simultaneously more simplified, organized, and unified.

This is the role Euler played in eighteenth-century mathematics.

At the time, mathematics had two thriving, well-developed neighborhoods, geometry and algebra. *Geometry* is the study of points, lines, planes, and the properties of figures built from them. It had been systematized in antiquity by Euclid's *Elements* (ca. 300 BC). One subdivision of geometry is trigonometry, concerned with the study of the relationships between the angles and lengths of sides in triangles, first developed as a tool of astronomy. *Algebra* is the study of equations with finite elements and discrete solutions, and largely concerned with rational numbers—numbers that can be expressed as integers or ratios of integers (in the form p/q) or, in what amounts to the same thing, numbers whose decimal representations repeat themselves. (Numbers like π, where the decimal values go on forever without repeating themselves, are said to be irrational.) Algebra had been largely organized, and given its name, in the Middle Ages by the Arab mathematician Mohammed ibn Musa al-Khowârizmî (ca. 780–850), thanks to his book *Hisâb al-jabr wa'l muquâbalah* (830). *Al-jabr* was al-Khowârizmî's term for the process of adding equal quantities to both sides of an equation to simplify it; after the word was transliterated into Latin as "algebra," it became the label for the entire field.

Unifying the Neighborhoods

By the beginning of the eighteenth century, mathematics was evolving a new neighborhood called *analysis,* or the study of—the collection of techniques for dealing with—infinities, for instance, series that include infinitely many numbers of terms. Analysis grew largely out of calculus, the study of continuous processes, which was developed by Gottfried Leibniz and by Newton (who called it the theory of fluxions). Analysis also involved the study of irrational numbers. And analysis treated imaginary numbers, or the square roots of negative numbers. These had been named by the philosopher and mathematician René Descartes, who seems to have thought them fictional—and his term stuck even as their uses and value to mathematics grew.

But it was Euler who organized analysis as a coherent body of knowledge, and transformed it into a thriving and organized area of mathematics: For instance, he carried out the first systematic study of functions. Functions are now-indispensable mathematical tools that pair or match one number with another (simple illustrations are formulas for calculating taxes, or for converting temperatures from Fahrenheit to centigrade). Euler also developed and expanded the tools that mathematicians had for summing infinite series of terms. Before him, mathematicians regarded summing infinite series of terms as an unpleasant duty that they sometimes had to do to solve problems when no other methods were available. But Euler showed that mathematicians need not be afraid of such series—they could be easy to work with, provided that the series converged. Euler was also the single most influential developer of mathematics notation in its history. Key symbols that he either introduced or standardized include:

> π, for the ratio of a circumference to the diameter of a circle, perhaps named after the first letter of the Greek word for "perimeter."
>
> e, for the base of natural logarithms, probably named for the first letter of "exponential"; the logarithm is the power to which a base must be raised to get a certain number, and e is the base of natural logarithms ($\log_e y = x$ means $e^x = y$).[4]

i, for $\sqrt{-1}$, the basic "imaginary number," which is hardly fictitious as Descartes thought but extends the range of equations that can be solved.[5]

$f(x)$, for a function of *x*, a function being a pairing or matching of one series of numbers with another.

sin, as an abbreviation for the sine function, pairing the measure of an angle in a right triangle with the ratio of the length of the opposite side to the hypotenuse.

cos, as an abbreviation for the cosine function, pairing the measure of an angle in a right triangle with the ratio of the length of the adjacent side to the hypotenuse.

$\sum$, for the summation of a series of terms.

In 1741, after 14 years in St. Petersburg, Euler left for the Berlin Academy at the invitation of Frederick the Great, another Renaissance man, though Euler remained in close correspondence with his St. Petersburg colleagues. Euler found Berlin less congenial than St. Petersburg. Frederick the Great was accustomed to highbrows with more flair than the taciturn Euler, thought him an aberration among his collection of pundits, and called him a "mathematical Cyclops."[6] In 1766, after 15 years in Berlin, Euler returned to St. Petersburg at the invitation of Catherine II, Catherine the Great. Though he was well supported, his health woes increased. He learned that he had a growing cataract in his remaining eye that would ultimately lead to blindness. He gamely coped—"I'll have fewer distractions," he remarked—learned to write with chalk on a slateboard, and taught his children to copy his calculations. Fewer distractions indeed. Euler pressed on undaunted for 17 years more, calculating, revising, composing, talking while walking around a table, with sons and assistants copying down his words. In this way, completely blind, he produced almost half of his entire oeuvre.

In 1771, a fire destroyed much of St. Petersburg. With his house burning, Euler himself—weak and blind—was carried to safety on the shoulders of a friend. He calculated on. On September 18, 1783, he tutored one of his grandchildren in math, worked out some problems regarding the paths of hot-air balloons, and considered possible orbits of the recently discovered planet Uranus, when his pipe suddenly dropped from his mouth. In the same breath, "he ceased to calculate and live."[7]

Today, the modern metropolis of mathematics is much larger still than it was in Euler's time. It is now laid out in huge boroughs, including analysis, algebra, and topology. Euler helped advance all three. His textbook on algebra, *Vollständige Anleitung zur Algebra* (Complete Instruction in Algebra, published in English as *Elements of Algebra*), presents the field essentially in the form it is today. He also made some of the first forays into topology, a field that did not exist yet, thanks to his famous solution to the Konigsberg bridge problem, involving the question of whether you could cross the seven bridges spanning the banks and islands of that city in a single walk without crossing any bridge twice—although topology would not be recognized as a borough for a hundred years or so.

But Euler was known as the master architect of analysis: scholars often called him "analysis incarnate." His most important single work in this field is a two-volume textbook, written during his Berlin years, entitled *Introductio ad analysin infinitorum* (*Introduction to Infinite Analysis*, 1748). In it, Euler unveiled numerous discoveries about functions involving infinite series, supplied proofs of theorems that others had left missing or incomplete, simplified many mathematical techniques, and proposed definitions and symbols that have since become standard, including π and e. "The *Introductio* did for analysis what Euclid's *Elements* had done for geometry and al-Khowârizmî's *Hisâb al-jabr wa'l muquâbalah* for algebra. It was a classic text from which whole generations were inspired to learn their analysis, especially their knowledge of infinite series."[8]

But the *Introductio* did much more than reorganize analysis. By translating many mathematical terms and expressions into the language of infinite series, it transformed analysis from a newly developing area of mathematics, alongside the existing fields of geometry and algebra, into its principal region. It all but made analysis the center city of mathematics.

The Deep Link

In the *Introductio*, Euler announced the dramatic discovery of a deep connection between exponential functions, trigonometric functions, and imaginary numbers. The proof grew out of his studies of exponential functions. In simplest terms, an exponential function involves a number called the base and another number set to the upper right of the base, called the exponent, with the exponent indicating how many times the base is multiplied by itself to produce the value of the function (this notation was invented by Descartes). A simple example of an exponential function is $y = 2^x$, where 2 is the base and x the exponent. For any integer x, this gives rise to a finite series of terms and an integer product. For instance, $2^2 = 2 \times 2 = 4$, $2^3 = 2 \times 2 \times 2 = 8$, $2^4 = 2 \times 2 \times 2 \times 2 = 16$, and so forth.

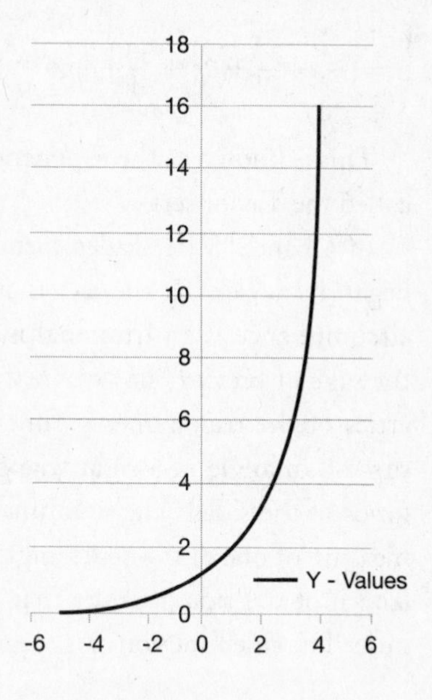

These integer pairs of numbers can be treated as belonging to a curve. On the infinite number of points on such a curve, only a few pairs are integers; the values on the dotted curve in between include decimals like 3.81 and even irrational numbers like $\sqrt{2}$ and π. What does it mean to multiply a number such as 2 by itself 2.31 or $\sqrt{2}$ or π times? For rational numbers expressible in the form p/q, this had to mean the qth root of 2 to the p. For example, 2 to the power 3.81 (= 381/100) would be the 100th root of 2 to the power 381. For irrational numbers, it would mean filling in the missing point on that curve, which can be calculated as the limit of an infinite sequence. Thus 2^{π} is the limit of 2^3, $2^{3.1}$, $2^{3.14}$, . . . , $2^{3.1415926}$, . . . as we take more and more decimal values of π.

In Chapter VII of the *Introductio*, Euler showed that, in choosing the base for an exponential function, there were numerous mathematical advantages to selecting the number created by adding up the following infinite series:

$$1 + 1 + \frac{1}{2!} + \frac{1}{3!} + \frac{1}{4!} + \frac{1}{5!} + \frac{1}{6!} + \frac{1}{7!} + \frac{1}{8!} + \frac{1}{9!} + \frac{1}{10!} + \frac{1}{11!} + \frac{1}{12!} \cdots$$

The sum of these terms, Euler noted, is the irrational number 2.718281828459 . . . , which "for the sake of brevity" he will represent as e. This number is the base of natural logarithms and one of the most important mathematical constants. Euler then noted that if we use e as the base of our exponents, then the function e^x can be calculated for any x by an infinite series:

$$e^x = 1 + x + \frac{x^2}{2!} + \frac{x^3}{3!} + \frac{x^4}{4!} + \frac{x^5}{5!} + \frac{x^6}{6!} + \frac{x^7}{7!} + \frac{x^8}{8!} + \frac{x^9}{9!} + \frac{x^{10}}{10!} + \frac{x^{11}}{11!} + \frac{x^{12}}{12!} \cdots$$

This is known as the exponential function, an example of what is called the Taylor series.[9]

In Chapter VIII, Euler turned to trigonometric functions. He began by reviewing the fact that if the diameter of a circle is 1 its circumference is an irrational number, 3.14159265 . . . which "for the sake of brevity" he says he will call π. He then described properties of the trigonometric functions, which associate to the measure of an angle in a right triangle the numbers created by various ratios of the sides. The sine function, for instance, associates to the measure of one of the acute angles in a right triangle the ratio of the length of the side opposite that angle to the length of the hypotenuse. The sine function can be generalized from acute angles to arbitrary angles as follows: Draw a right triangle *ABC*, with hypotenuse *BC* of length 1, in the (*x*, *y*)-plane so that vertex *B* lies at the origin (0,0), vertex *A* lies on the positive *x*-axis, and vertex *C* lies above the *x*-axis. Let *a* be the measure of the angle ∠ *ABC*, measured counterclockwise from the positive *x*-axis. Then sin *a* is the ratio of the lengths *AC*/*BC*, but since *BC* = 1, sin *a* = length *AC* = the *y*-coordinate of the point *C*. If we take "*y*-coordinate of *C*" as the definition for sin *a* (*a* is the measure of angle ∠*ABC*), then we have a definition that works for any angle: rotate *BC* through an angle *a* (starting from the positive *x*-axis and measured counterclockwise) and record the *y*-coordinate of *C*. Then the sine goes from 0 to 1 (at 90 degrees) back to 0 (180 degrees), thence to −1 (270 degrees), again to 0 (360 degrees), and repeats that pattern through successive cycles of 360 degrees, producing a pattern called the "sine wave" familiar from oscilloscopes. The general cosine function is defined in the same way, except taking the *x*-coordinate of *C* as the value. As the angle varies, the cosine goes from 1 to 0 to −1 to 0 to 1, repeating just as does the sine function, making the same pattern as the sine function, but out of phase.

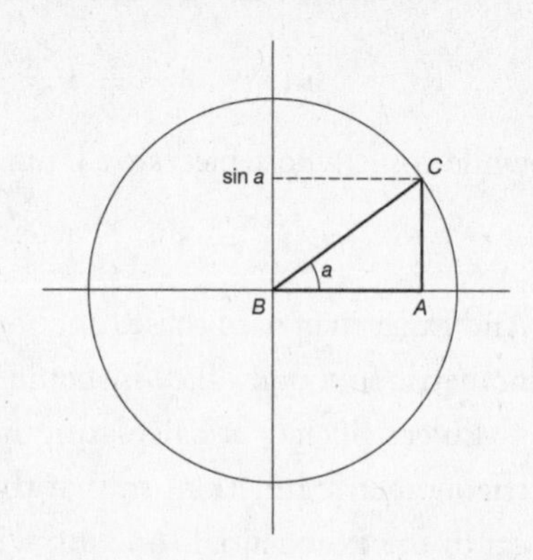

Euler then runs through various more or less obvious properties of sines and cosines, including the fact that, from a simple application of the Pythagorean theorem, $(\sin x)^2 + (\cos x)^2 = 1$.

Continuing to summarize things that Newton and other predecessors knew, Euler next showed how trigonometric functions involving sines and cosines could also be expressed in terms of infinite series. For instance, the function sin *x* can be expressed as the following infinite sum of terms, which we'll put in this font so that we can simply and easily follow the terms:

$$\sin x = x - \frac{x^3}{3!} + \frac{x^5}{5!} - \frac{x^7}{7!} + \frac{x^9}{9!} - \frac{x^{11}}{11!} \cdots$$

while cos x, whose pieces we'll put in **this font**, is:

$$\cos x = \mathbf{1} - \frac{\mathbf{x^2}}{\mathbf{2!}} + \frac{\mathbf{x^4}}{\mathbf{4!}} - \frac{\mathbf{x^6}}{\mathbf{6!}} + \frac{\mathbf{x^8}}{\mathbf{8!}} - \frac{\mathbf{x^{10}}}{\mathbf{10!}} \cdots$$

And Euler then used these functions to show how all the other trigonometric functions likewise could be expressed as infinite series.

Euler's fluency at calculating now enabled him to arrange these trigonometric functions so that they added up to something identical to the exponential function with base *e*. He did so with the aid of the imaginary number $\sqrt{-1}$, which he would later—years after he wrote the *Introductio*—symbolize as *i*. Although *i* is not a "real" number—a number with a place on a number line—it is used in real mathematical operations and allows mathematicians to solve otherwise insoluble equations. If, for instance, you insert it in the exponent of e^x it shows up in each term of the infinite series associated with it:

$$e^{ix} = \mathbf{1} + ix + \frac{(i\mathbf{x})^2}{\mathbf{2!}} + \frac{(ix)^3}{3!} + \frac{(i\mathbf{x})^4}{\mathbf{4!}} + \frac{(ix)^5}{5!} + \frac{(i\mathbf{x})^6}{\mathbf{6!}} + \frac{(ix)^7}{7!} + \frac{(i\mathbf{x})^8}{\mathbf{8!}} \cdots$$

But i^2 is -1, and therefore $i^3 = -i$, $i^4 = 1$, $i^5 = i$, etc. So the series becomes:

$$e^{ix} = \mathbf{1} + ix - \frac{\mathbf{x^2}}{\mathbf{2!}} - \frac{ix^3}{3!} + \frac{\mathbf{x^4}}{\mathbf{4!}} + \frac{ix^5}{5!} - \frac{\mathbf{x^6}}{\mathbf{6!}} - \frac{ix^7}{7!} + \frac{\mathbf{x^8}}{\mathbf{8!}} \cdots$$

Euler observed that if you group together the multiples of *i*, you obtain:

$$e^{ix} = (\mathbf{1} - \frac{\mathbf{x^2}}{\mathbf{2!}} + \frac{\mathbf{x^4}}{\mathbf{4!}} - \frac{\mathbf{x^6}}{\mathbf{6!}} + \ldots) + i(x - \frac{x^3}{3!} + \frac{x^5}{5!} - \frac{x^7}{7!} + \ldots)$$

Or, as he wrote toward the end of Chapter VIII of the *Introductio* (using *i* where he used $\sqrt{-1}$, as does the English translation),[10]

$$e^{ix} = \cos x + i \sin x$$

This equation establishes the deep connection between exponential and trigonometric functions. When the great Indian mathematician Srinivasa Ramanujan (1887–1920) discovered this connection on his own while in high school, he wrote it down excitedly—and was so crestfallen to discover that he was not the first that he hid all his calculations in the roof of his house.[11]

This equation is magic enough, but there's more. Suppose x is π. The sine of π is 0, and the cosine of π is -1. Then $e^{i\pi} = -1$, or $e^{i\pi} + 1 = 0$.

Another way to show the truth of this equation graphically is the following. Suppose we insert π for x. Then the formula a few paragraphs above becomes:

$$e^{i\pi} = 1 + i\pi + \frac{(i\pi)^2}{2!} + \frac{(i\pi)^3}{3!} + \frac{(i\pi)^4}{4!} + \frac{(i\pi)^5}{5!} + \frac{(i\pi)^6}{6!} + \frac{(i\pi)^7}{7!} + \frac{(i\pi)^8}{8!} \cdots$$

Mathematicians can add such a sequence as a series of vectors, with each one beginning at the tail end of the one preceding it, and with the imaginary number i rotating a vector 90 degrees counterclockwise.[12] If we start at 0, the first term (1) is a vector that takes us 1 unit out on the x-axis, to coordinate (1, 0). The second term ($i\pi$) takes the form of a vector that starts at (1, 0) and, rotated counterclockwise with respect to the first, goes straight up π units, ending at coordinate (1, π). The third term, ($i\pi^2/2!$), takes the form of a vector that starts at (1, π) and—rotated another 90 degrees from the previous one—runs in the opposite direction from the first, going across the $y = 0$ line to the point $(-(\pi^2/2 - 1), \pi)$. The fourth term is a vector that runs downward, ending up below the x-axis, and so forth. Because the vectors keep rotating 90 degrees counterclockwise, and keep getting shorter because the denominator increases much faster than the numerator, the result is a polygonal spiral that converges on the point $(-1, 0)$ (see diagram on next page).

Euler's simple formula (according to some definitions, it is not an equation in this form, for it contains no variables) contains five of the most fundamental concepts of mathematics—zero, one, the base of the natural logarithms e, the imaginary number i, and π—as well as four operators—addition, multiplication, exponentiation, and equality—and each exactly once. It states that an irrational number multiplied by itself an imaginary number times an irratio-

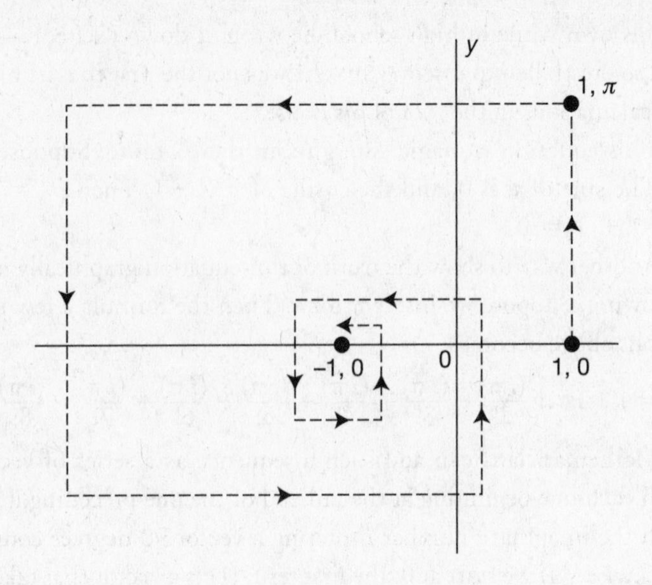

Polygonal spiral, showing how the infinite series converges to −1.

nal number of times—plus one—equals exactly zero. The numbers π^e, 2^π, and e^π are all thought to be irrational. But $e^{i\pi}$ picks out that special place in the architecture of numbers where rational, irrational, and imaginary numbers mix in a way that spookily "balances out" to exactly zero. It has been said that all analysis is centered here in this equation.[13] Among other things, Euler's result here demonstrated that imaginary numbers, despite Descartes' scorn, were not on the margins, but at the very center of mathematics. They would play a greater and greater role in mathematics—and then, with the advent of quantum mechanics in the twentieth century, in physics and engineering and any field that deals with cyclical phenomena such as waves that can be represented by complex numbers. For a complex number allows you to represent two processes such as phase and wavelength simultaneously—and a complex exponential allows you to map a straight line onto a circle in a complex plane.

It may be true, then, that Euler's formula $e^{i\pi} + 1 = 0$ is but one implication, one step, in his exploration of functions. It is "only" an equation, a single one of the thousands of steps in the ongoing process of scientific inquiry, a mere implication in Euler's extended exploration of functions. And yet, some steps in an inquiry acquire, and deserve, special status. Certain expressions serve as landmarks in the vital and bustling metropolis of science, a city that is continually undergoing construction and renovation. They preserve the work of the past, orient the present, and point to the future. Theories, equipment, and people may change, but formulas and equations remain pretty much the same. They are guides for getting things done, tools for letting us design new instruments, and repositories for specialists to report and describe new discoveries. They summarize and store, anticipate and open up.

Euler's equation, too, emblematized the way that its author had recast mathematics. Mathematics, like other sciences, does not develop along a predetermined track, but follows a historically contingent path in which each generation of scientists inherits assumptions, techniques, and concepts from the generation before, transforms them, and passes them on in turn. Thanks to this process, we perceive the field as structured in a particular way, as having a certain ontology, with different phenomena assigned to distinct domains. Every equation implicitly refers to this inherited structure. But Euler rearranged this ontology, reorganizing it so that analysis was at the center, with geometry and algebra as neighborhoods. Looking backward, mathematicians may take the latest organization as self-evident—which, no doubt, is why the mathematician Carl Friedrich Gauss is said to have remarked that anyone to whom $e^{i\pi} + 1 = 0$ is not obvious is not a mathematician. When you are fully literate, nothing comes as a surprise. But mathematicians are made not born; in infancy they are not yet mathematicians, and have to learn it—and in such learning often experience extensive transformations and reorganizations of mathematical knowledge that they have only partially acquired. The brief formula $e^{i\pi} + 1 = 0$ is the most succinct expression of this process.[14]

There is yet one more, still deeper reason why this formula has become an icon. As Devlin once wrote of Euler's equation, "Euler's equation reaches down into the very depths of existence. It brings together mental abstractions having their origins in very different aspects of our lives, reminding us once again that things that connect and bind together are ultimately more important, more valuable, and more beautiful than things that separate."

Devlin's remark suggests the chief reason why an equation such as Euler's attracts value and interest beyond the particular scientific inquiries that gave birth to it. It serves as a clear and concise example of what an equation and formula does: it shows how what seemed to be disparate and even incompatible elements (rational, irrational, and imaginary numbers) are implicated in a unity, and does so concisely, with few moving parts, so to speak. It simultaneously simplifies, organizes, and unifies. It brings what equations do out into the open. It is an equation that shows what it is to be an Equation.

INTERLUDE

Equations as Icons

Journalist: Do the Russians have anything like GISMO?
Scientist (Rod Taylor): No, I'm sure they'd like it, though.
Journalist: Can you give us the equation?
Scientist: No, I'm sure they'd like that even better.

—*The Glass Bottom Boat* (1966)

Equations have a subtle influence on the fabric of our language and our thought far beyond science. Cloaking thoughts in mathematical dress seems to make them more authoritative, certain, precise, and eternal. Jokes, maxims, political bumper stickers, and uplifting self-help slogans are often revamped as equations: "Knowledge = Power," "War = Killing People," "Preparation + Patience = Success." Equations are written *about* humor, as the following diagnosis of the 2007 episode in which CBS Radio fired talk show host Don Imus for a racist remark: "White guy plus black slang equals comedy. But there's where the equation breaks down. White guy plus black slang minus common sense equals tragedy."[1]

Or consider George Orwell's famous equations from his novel *1984*, which, though obviously overtly false, point to a different kind of truth:

War = peace
Ignorance = strength
Freedom = slavery

While these have a superficial resemblance to equations of math and science, they are really just metaphors in disguise. The "=" sign in them does not mean the mathematical notion of "equal to" or even equivalence. In mathematics, "=" is quantitative, and means "is exactly the same as," referring to the number of items in a set, or to a specific measurable amount. This is the foundation stone of the discipline. The way that knowledge is power, to take one example, is qualitative and much different, and must be addressed by broaching the philosophical complexities of the meanings of the seemingly self-evident words "sameness," "equality," and "is."

Still, these fanciful equations are intriguing, for they exhibit the dangerous hope that other kinds of knowledge can be couched in equational terms, with neat packaging, balanced amounts, and simple units. Equations, that is, can seduce us into thinking that this is the way to think, and that other ways are inferior or even defective. A correspondent to a science magazine, after receiving an email from someone at a breakfast cereal company asking him to produce an equation for the best time to add the milk, made light of the way the public seems obsessed with finding equations for even the most trivial of actions. The letter unleashed comments from others—who had seen requests for equations for making sandwiches, parking cars, and "the perfect sitcom"—warning that the practice had a dark side, not only because it was bad science, but because it encouraged irresponsible behavior among scientists and mistaken views about the nature of science among the public.[2]

Specific equations, too, can have a wide range of symbolic meanings. Take 2 + 2 = 4, the slightly elder sibling of 1 + 1 = 2. In fiction and reality, it has been used to symbolize the superiority of the irrational over the rational, the rational over the irrational, and the divine over both the rational and the irrational.[3] In Dostoyevsky's novel *Notes from Underground*, for instance, the narrator describes it as "insufferable," as a "piece of insolence," as sterile and rational, as something dead and beneath bare consciousness, which the narrator finds is "infinitely superior to two times two makes four." In George Orwell's novel *1984*, on the other hand, Winston, the protagonist, uses $2 + 2 = 4$ as a self-evident truth, the touchstone of sanity and rationality, available to thinking at any and every moment, the one shining light to grasp that objective reality, which assures and even guarantees for him that objective reality is there; for the Party, $2 + 2 = 4$ is the final resistance that must be defeated in the way of the success of doublethink and the Party's rule, the one outside standard that must be eradicated. Orwell, in turn, was only quoting a genuine slogan by the leaders of the Soviet Union, which used $2 + 2 = 5$, written on billboards and in electric lights, as a symbol of optimism, of the ability of labor to triumph over nature, of the fact that "miracles could be worked through the sorcery of naked force."[4] The correct equation—dry, rational, and stale—was false, it seems, because it did not capture human creative ability, while the incorrect one was true because it symbolized the way human creative ability can overcome limitations of nature. Meanwhile, the architect and inventor Buckminster Fuller liked to define synergy with the motto "$1 + 1 = 4$," meaning that efficient and imaginative use of parts produces more than is possible with conventional methods. Finally, the eminent Oxford theologian Marilyn McCord Adams, arguing that "Human nature is not created to function independently, but in omnipresent partnership with its Maker," speaks of the "self-effacing Spirit" who "is ever the midwife of creative insight, subtly nudging, suggesting, directing, directing our attention until we leap to the discovery that $2 + 2 = 5$."[5]

Novelists who have used equations in bizarre ways include Italo Calvino, whose book *Cosmicomics* features Einstein's general relativity in one story. Another is Mark Leyner, whose book *Et Tu, Babe* includes a character who claims to have tattooed on his penis Max Planck's energy formula $E = h\nu$, hence, something associated with radiation and power—and is humiliated to have to confess in front of a judge that it is actually $d = 16t^2$—Galileo's law of falling bodies.

If equations have a dark side, it is that they can also tempt us to think that knowledge resides in the equation itself, rather than in the ongoing construction and renovation of the city of science (what Plato called more questioning). They can promote the erroneous view that science consists of a set of facts or beliefs to be memorized, rather than a quest for greater understanding to be achieved by moving beyond the facts or beliefs we already have.

CHAPTER FIVE

The Scientific Equivalent of Shakespeare: The Second Law of Thermodynamics

$$S' - S \geq 0$$

DESCRIPTION: The entropy of the world strives toward a maximum.
DISCOVERER: An international cast of characters
DATE: 1840s–1850s

> A good many times I have been present at gatherings of people who, by the standards of the traditional culture, are thought highly educated and who have with considerable gusto been expressing their incredulity at the illiteracy of scientists. Once or twice I have been provoked and have asked the company how many of them could describe the Second Law of Thermodynamics. The response was cold: it was also negative. Yet I was asking something which is about the scientific equivalent of: *Have you read a work of Shakespeare's?*
>
> —C. P. Snow, *The Two Cultures*

The first two laws of thermodynamics are easy to state. Rudolf Clausius, who formulated the second one and who coined the word "entropy" as a name for a measure of disorder, expressed them this way: "The energy of the world is constant; the entropy of the world strives toward a maximum." A popular formulation in simple language is: "You can't win. You can't break even, either." Max Planck's symbolic formulation of the second law, with S the entropy at an earlier time and S' the entropy at a later time, is given above.

ABOUT THE CHARACTERS

Ludwig Boltzmann (1844–1906)

Austrian physicist tormented by depression and mood swings. Uses statistical methods to show how agitations of swarms of tiny atoms give rise to bulk properties of matter. Famous for the Boltzmann equation, $S = k \log W$. Deeply troubled by attacks on atomic theory, and thus his work, by prominent colleagues. In 1906, on vacation near Trieste, hangs himself while his wife and daughter are swimming. His equation is engraved on his tombstone in Vienna.

Rudolf Clausius (1822–1888)

German physicist who settles the battle between the conversionists and the conservationists by declaring that *two* principles are in play, one involving the *conservation* of what is soon called energy in exchanges of heat and mechanical work, the other the *conversion* of heat into energy. Coins the word "entropy," which he refers to as *S*, and is another claimant as discoverer of the second law.

Lazare Carnot (1753–1823)

French military engineer specializing in the elimination of waste. Military duties interrupt his studies of inefficiency in water-powered machines. Does a jail spell when he seduces a woman betrothed to another man. Nicknamed "Organizer of Victory" by French Revolutionaries thanks to his creative problem-solving for the cause. Fathers and homeschools two sons: Sadi and Hippolyte.

Hermann von Helmholtz (1821–1894)

German physicist who masters and contributes to an astounding variety of fields, including acoustics, aesthetics, anatomy, biology, magnetism, mathematics, mechanics, meteorology, ophthalmology, optics, phenomenology, philosophy, physics, physiology, and psychology. Invents the ophthalmoscope for examining the inner eye. Mentors many scientific stars, including Nobel Prize winners Albert Michelson, Max Planck, and Wilhelm Wien.

Sadi Carnot (1796–1832)

Son of Lazare, a quiet engineer who inherits from his father an apartment and an interest in reducing inefficiency in heat engines. Writes the most seminal work on the subject, *Reflections on the Motive Power of Heat*, which his brother Hippolyte edits. It contains the notion of conservation, reversibility, and the famous "Carnot cycle." But the book is ignored, Carnot stops publishing, catches scarlet fever, brain fever, cholera, and dies, age 36, in a madhouse.

James Prescott Joule (1818–1889)

As a youth, James builds a home science lab in his parents' brewery. A few years later, he manages highly accurate measurements of various conversions of heat and electrical, mechanical, and chemical energy into one another. He's the first to measure the mechanical equivalent of heat. His work promotes the idea of the conversion of energy, and sets off a battle between proponents of conversion and of conservation.

James Clerk Maxwell (1831–1879)

An improbable prodigy taunted by cruel classmates who nickname him "Dafty" for his plain clothes, country accent, and candid questions. Establishes the field of electromagnetism via one of the most brilliant uses of analogy in history, and lays the groundwork for the electronic age. Explains, among other things, the rings of Saturn, the behavior of gases, and the nature of spinning tops, constructing "the fanciest top ever made." Dies at age 48.

Count Rumford (1753–1814)

British soldier of fortune, amateur scientist, and spy, who conducts experiments on heat in between courtships of wealthy widows. Refutes the "caloric" theory of heat proposed by the former husband of his latest conquest. Proclaims that heat is not a substance but comes from motion generated by friction, and uses this idea to quantitatively compare different kinds of work. Thinks he's another Newton.

Robert Mayer (1814–1878)

While a doctor on a boat in the East Indies, notices the unusual redness of his crew's blood, meaning it is oxygen-rich. Deduces that human metabolism is slower in the tropics and that mechanical work and heat are interchangeable. His unintelligible paper on the subject is rejected by a journal, though he later revises and publishes it. Depressed by rejection of his claim to the second law, he flings himself from a third-floor window and is sent to an asylum in a straitjacket.

William Thomson (1824–1907)

A polymathic, trilingual, and farsighted son of a mathematics professor, the future Lord Kelvin. He's torn by the conflict between the conversionists and the conservationists, and is determined to make peace. Developer of the new science of heat-mechanics, which he names *thermodynamics*. Co-author of thermodynamics' *Principia*, the *Treatise on Natural Philosophy*, and one of several claimants as discoverer of the second law.

Max Planck (1858–1947)

Undeterred by his professor's warning that everything in physics has been discovered already, while focusing on neatening up the old—tidying up thermodynamics— he invents the quantum and changes the world! His eldest son dies in World War I at Verdun; his second eldest is hanged in World War II for joining the plot to kill Hitler. Wins the Nobel Prize in 1919. A world-famous research organization is named for him. So is a 43-kilometer-long asteroid.

Wilhelm Wien (1864–1928)

A farmer at heart, takes on physics as a second career. Authors Wien's law, which uses the second law of thermodynamics to map radiation's dependence on temperature, thereby leading us "to the very gates of quantum physics." Discovers a positively charged particle which, when further explored by others, becomes the proton. Wins the Nobel Prize in 1911. A crater on Mars, 120 kilometers in diameter, is named after him.

This law is essential to the activities of the world around us. If you do not understand this, you can have little understanding of how the world works. This was surely C. P. Snow's motivation in saying that asking if someone can describe the second law of thermodynamics is like asking, "Have you read a work of Shakespeare's?" It should be equally shameful for people who think themselves cultured to have to answer no to either question.

My thoughts on this subject are even more radical. I think that the second law of thermodynamics is actually Shakespearean. Its story involves powerful and finely drawn characters. It has fundamental implications for human life. And it unfolded in somewhat the way Shakespearean dramas do.

Here's a plot summary of how one version might go.

PROLOGUE

Europe, end of the eighteenth century

A new mechanics is on the horizon. The steam engine and other technologies have drawn attention to phenomena relating to heat. Driven by practical necessity and curiosity, legions of inventors are attempting to develop better steam engines. But their work is mostly tinkering, because as yet little is known about heat. Heat seems to be a force—we can put it to work!—but not one whose operations are explained by Newtonian pushes and pulls. A theory of heat is clearly needed, and a crude one, called "caloric theory," appears. Developed in the second half of the eighteenth century by French scientist Antoine Lavoisier—the "father of modern chemistry"—caloric theory conceives of heat as an invisible and weightless fluid that flows from place to place, which provides the beginning point for understanding heat as a force. Several scientists, whose motives range from curiosity and professional duty to pride and ambition, turn their attention to this heat-force. They are soon embroiled in a conflict about whether utilizing this heat-force involves conservation or conversion of heat: is the total amount of heat always the same, or does it get converted to something else? The resolution of this conflict will turn out to be the key to the new mechanics.

ACT ONE

Paris and Munich, end of the eighteenth century

Scene 1. Paris, 1803

Lazare Carnot (1753–1823), a military engineer whose talent is uncovering and eliminating administrative and mechanical inefficiency, publishes a treatise on water-powered machines, *General Principles of Equilibrium and Motion*. Follow the water, he writes: the maximum power depends on how great a distance it falls. Track down and eradicate sources of waste, he also counsels, to make your machine work better. But Lazare can't pursue these insights. He's forced back to military duties; later he seduces a woman betrothed to another and ends up in jail. He's released as the French Revolution begins and joins the revolutionaries, who nickname him "Organizer of Victory" for the innovative way he mobilizes, trains, and supplies troops. He has two sons, whom he homeschools and who will carry on his legacy: Sadi, a military engineer (named after a Persian poet), and Hippolyte, a journalist and politician.

Scene 2. Munich, 1797–98

Count Rumford (1753–1814), soldier of fortune and amateur scientist, is in Munich, momentarily between courtships of wealthy widows. Keen to reveal the mysteries of heat, he puts a 6-pound brass cannon barrel in a vat of water, inserts a drill bit driven by a winch, hitches up a horse to the winch, and finds that this generates enough heat through the drilling to boil the water in 2½ hours. The caloric theory formulated by Lavoisier (the former husband of one of Rumford's mistresses) is wrong, Rumford proclaims, for the seemingly inexhaustible amount of heat generated in the process is not coming from either the brass or the water, but is clearly a form of motion coming from the friction between the bit and the cannon. He counts the candles it takes to boil the same amount of water, to compare the amount of heat and mechanical force. Reporting to the Royal Society, he implicitly likens himself to Newton, saying the laws of heat are as important as those of gravity. But Rumford is no Newton. His arguments are not entirely convincing and he has no overall theory of heat, just a very suggestive set of observations. Yet his idea that one can quantitatively compare various kinds of work that create the same amount of heat (candles, horses), and the work that heat does in different forms, helps to set up the looming conflict between conservation and conversion.

ACT TWO

Paris, Manchester, and Oxford, 1820s–40s

Scene 1. Paris, 1823

Sadi Carnot (1796–1832), a quiet engineer, returns from his father Lazare's deathbed to the apartment he has just inherited. Determined to carry on his father's work, Sadi sets to work composing a treatise, *Reflections on the Motive Power of Heat*, on ways to make steam engines more practical and efficient. Fearing that his prose is too convoluted to appeal to the general audience that he covets, he has his brother Hippolyte edit his manuscript and steady his prose. Steam engines, he begins, "seem destined to produce a great revolution in the civilized world." Nevertheless, he continues, "their theory is very little understood." Such a theory must begin by considering the general question of what the most efficient way is to use steam. One key thing to consider in a machine, Carnot realizes, is its maximum duty, or maximum output; for instance, how high a given amount of coal in the machine can raise a given amount of water. Follow the heat, he writes. Caloric in a heat engine, like water in a water engine, is conserved as it flows from the hot to the cold places, and the maximum power depends on the magnitude of the temperature drop. The most efficient machine is modeled by an ideal cycle of expansion and compression in which the engine works reversibly, the caloric being conserved in going back and forth between the two temperature endpoints with no heat diverted (wasted) to friction or dissipation. This is a key insight, but *Reflections* is almost totally ignored. Carnot publishes nothing more, catches scarlet fever, brain fever, cholera, and dies, aged thirty-six, in a madhouse.

Scene 2. Manchester, 1840s

James Prescott Joule (1818–1889), who as a youth built a home lab in his parents' brewery, manages to get highly accurate measurements of various conversions of heat and electrical, mechanical, and chemical energy into each other; for instance, the temperature increase that rotating paddles, stirring up water, produce in the water thanks to friction. He determines the mechanical equivalent of heat: 772 foot-pounds of work make a 1-degree F rise in 29 cubic inches of water.

Scene 3. Oxford, 1847

The conflict between conservation (Carnot's approach) and conversion (Joule's) begins to come to a head. Young William Thomson (later Lord Kelvin, 1824–1907), the polymathic, trilingual, and far-sighted son of a mathematics professor, travels to Paris, where he reads the only published comment on Sadi Carnot's work and is so impressed he tries in vain to find a copy of the original. Then he attends a conference in Oxford, where he hears Joule. Joule is treated badly by the conference organizers, who instruct him to be brief. But Joule's words jolt Thomson. How can heat be converted to something else when Carnot's spectacular work relies on the fact that the amount of caloric in an engine is constant? Joule's work must have "great flaws," Thomson decides, and he resolves to find them.

ACT THREE

Great Britain and Germany, 1840s–60s

Scene 1. Glasgow

Thomson, still convinced that Carnot's conservation theory is right and that something must be wrong with Joule's work, gets another jolt. He reads a paper by German physicist Rudolf Clausius (1822–1888), who has also noticed the conflict between the approaches of Carnot and Joule. Clausius has been examining the kinetic theory according to which heat and gases consist of tiny particles in constant motion. And Clausius says that the conflict between Carnot and Joule is only apparent, and is not a conflict in reality because *two* principles are in play. One involves the *conservation* of something (not heat, and soon called energy) in exchanges of heat and mechanical work; the other the *conversion* of heat into energy, and the property that heat cannot flow spontaneously from colder to warmer bodies. Thomson, inspired, begins to leapfrog works with Clausius on the new heat-mechanics. In 1854 Thomson names it *thermodynamics*, after the Greek for heat and force. Some heat in every engine, Thomson writes, "is irrevocably lost to man, and therefore 'wasted' although not *annihilated*"—his version of the second of Clausius's two principles. Clausius embarks on a series of papers that culminate in 1865, when he names the tendency of the energy transfer process to occur spontaneously (disorder, we now say) "entropy," after the Greek for "transformation"; he referred to entropy as S, a function of the state of a system, and uses the formula $\int dQ/T \leq 0$. In 1867, Thomson and his collaborator Tait compose thermodynamics' *Principia*, the *Treatise on Natural Philosophy*. In 1872, Clausius formulates what becomes known as the two laws of thermodynamics this way: "The energy of the world is constant; the entropy of the world strives toward a maximum."

Scene 2. Heilbronn, Germany

Priority battles erupt. In 1847, German physician Robert Mayer (1814–1878) reads a paper by Joule on the conversion of heat into mechanical energy and says he discovered it first. Seven years previously, as a doctor on a Dutch ship in the East Indies, Mayer had realized that the unusual redness of the blood of the crew—meaning it was oxygen-rich—was due to the fact that human metabolism is slower in the tropics. This had inspired him to write a paper on the interchangeability of mechanical work and heat to *Annalen der Physik und Chemie*, the leading German science journal, but the poorly written paper had been treated as a crackpot letter by the editor and Mayer didn't receive a reply, though he later revised it and published it elsewhere. Depressed when Joule disputes his priority, Mayer flings himself out a third-floor window, and is committed to an asylum in a straitjacket. Meanwhile, another German physicist, Hermann von Helmholtz (1821–1894), is also a contender for discovering the first law of thermodynamics thanks to an 1847 paper on "the conservation of force." Tait and Clausius battle over who discovered various principles of thermodynamics, slinging mud at each other in journals and books.

ACT FOUR

London, Graz, and Vienna, 1870s

Scene 1. London and Graz

Another battle breaks out, this time over which of the two laws of thermodynamics is more important. For they seem to conflict. The first law (conservation of heat/energy) implies that processes are reversible—that the "before" and "after" states of a physical process cannot be distinguished, for each can turn into the other. The second law (heat cannot be completely turned back into work) implies irreversibility, or what is later known as the "arrow of time," that change tends to go in one direction only. The problem comes to a head in Clausius's specialty, the kinetic theory of gases. A gas is a "big thing" governed by irreversible processes and the second law, but is composed of "little things"—atoms and molecules—each of which obeys reversible Newtonian principles governed by the first law. In 1859, James Clerk Maxwell (1831–1879) comes across Clausius's paper on the kinetic theory of gases, and decides that the statistical methods he had just used for studying Saturn's rings as a collection of small bodies might also apply to gases. Maxwell realizes that a gas's jostling molecules do not end up, or reach equilibrium, with all of them having exactly the same speed; rather, they have a range of speeds clustered about one value. Imagine a dense crowd of people randomly milling about in a train station: the people are not all moving at exactly the same speed, but most are moving at about the same speed, with only a handful dead-still or going very fast. To understand the behavior of the gas, furthermore—or of a crowd—it is not necessary to track the positions and velocities of each and every individual in it, but suffices to know the distribution of positions and velocities. Using only statistical methods and assumptions of Newtonian mechanics, Maxwell comes up with an equation to describe the spectrum of velocities of the gas molecules. The plot describes a bell-shaped curve: few lie at the extremes, moving almost not at all or very fast, but most cluster about an average velocity with the numbers tapering off at higher and lower velocities. But Clausius's 1865 paper, and Maxwell's own experimental work, force him to modify the theory, and he publishes a revision in 1867. Maxwell concludes that the second law is merely statistical, true only when vast quantities of particles are involved and not true of individual motions. The second law, he writes, is true for the same reason as is the statement that "if you throw a tumblerful of water into the sea, you cannot get the same tumblerful out again (i.e., exactly the same molecules as before)."[1] At the atomic level, reversibility is possible and the second law does not hold, he thought. But why can't reversibility be possible for large bodies in principle; why cannot heat flow sometimes from a cold to a hot body? In 1867, writing to Tait, Maxwell demonstrates this imaginatively and theatrically by a thought experiment involving a little "being" that can detect faster-moving molecules in a gas, and by opening and shutting a door at the right time gets the faster ones on one side of a barrier, thus causing heat to flow to one side of the box. In this way the creature seems to refute Thomson's idea of dissipation by getting heat to flow from a colder to a hotter place. Maxwell publishes this idea in 1871, in a short section entitled "Limitations of the Second Law of Thermodynamics" in his *Theory of Heat*. That seems to end the matter; it is all a question of statistics.

Scene 2. Graz and Vienna, 1870s

Ludwig Boltzmann (1844–1906) extends Maxwell's work. In 1868, a year after Maxwell's paper, he produces an expression for the distribution of energy among molecules of a gas valid for gases of any kind. To derive it, he relies on a key assumption known as the equipartition theorem, according to which a molecule stores energy by spreading it equally among all the avenues ("degrees of freedom") available to it. This work also includes a now-famous term, Boltzmann's constant, now referred to as k, 1.38×10^{-23} joules/Kelvin. The result is a thoroughly statistical interpretation of thermodynamics. In 1872, Boltzmann generalizes this work still further, in a revolutionary paper with a banal title, "Further Researches on the Thermal Equilibrium of Gas Molecules." In it, he derives a function related to entropy, now called the H-function, whose value almost always increases with time until it reaches a maximum—an entirely novel and innovative approach to proving the second law, and in a way that explicitly demonstrates irreversibility, or how it increases with time. But this work is subject to friendly fire: from Thomson, in an 1874 paper that refers to Maxwell's little creature as a "demon" and, in 1876, from Boltzmann's former mentor Josef Loschmidt, for not having eliminated certain puzzles involving the relation between the second law and the first. Even complex many-body systems, such as the deployment of the planets around the sun, are cyclical, with the same patterns eventually recurring, so why doesn't this happen in thermodynamic systems? Also, if two gases mixed, following the H-curve, entropy increases—but if you then reverse the velocities of all the gas molecules, wouldn't it then make the H-curve, the "arrow of time," go the other way, violating his theorem? Boltzmann replies (1877) that when a big state corresponds to many equally probable little states, its probability is related to the number of little states. This all but forces big states to evolve in the direction of their more probable states. Boltzmann's approach is an explicit probabilistic interpretation of entropy, introduces probability into electromagnetism, and proves the centrality of irreversibility to thermodynamics. Newton's laws + objects made of myriads of pieces + laws of probability = the arrow of time. The forbidden becomes the highly unlikely. On large scales, you play dice, and statistics rule. In 1879, this work is extended by Boltzmann's former teacher Stefan into the Stefan-Boltzmann law, a law relating the dependence of radiation on temperature in black bodies. But Boltzmann becomes vulnerable to depression late in life, from both personal and professional setbacks, and in 1906, on vacation near Trieste, he hangs himself while his wife and daughter are out swimming. On his tombstone is engraved his equation, not in the form he wrote it but as Max Planck would: $S = k \log W$.

ACT FIVE

Berlin, 1890s

Scene 1. Berlin, early 1890s

Physicist Wilhelm Wien (1864–1928), an introvert who is repeatedly thwarted in his attempts to become a farmer like his parents, extends Boltzmann's ideas about the second law of thermodynamics. He is at the Physikalisch-Technische Reichanstalt, the imperial bureau of standards, and he, along with several other scientists, is studying something called "black body radiation" out of a mixture of theoretical interest and practical motives (the calibration of electric lamps). If you heat up a body that absorbs all the radiation that falls on it (a "black body"), it eventually begins to glow and emit light itself. According to classical mechanics, something in the body is like a resonator, acting like a miniature antenna, in picking up and giving off energy in the form of electromagnetic waves. It is as if the electric charges were located in springs of varying flexibilities, and they oscillated back and forth at a rate depending on the flexibility of the spring and with an intensity that depended on the temperature. Maxwell had fully explained the radiation's creation, absorption, and propagation. Experimenters had measured this radiation and produced curves of wavelengths and intensities at each temperature. Using the Stefan-Boltzmann work, Wien writes a paper entitled, "A New Relationship Between the Radiation From a Black Body and the Second Law of Thermodynamics." The paper contains Wien's law, which uses the second law to map radiation's dependence on temperature at high temperatures: "In the normal emission spectrum from a black body each wave-length is displaced with a change of temperature in such a way that the product of temperature and wave-length remains constant."[2] It is called a "displacement law," and is reformulated in 1896. That year, scientists at the bureau construct a special oven to measure the wavelengths of the radiation, focusing on the shorter wavelengths that are easier to detect. Wien's law says the energy emitted increases with temperature, though the increase is not equally distributed across all wavelengths but shifts toward shorter wavelengths. But experimenters discover that as the energy curve is extended to longer and longer wavelengths, Wien's law breaks down, though when the energy gets low enough, it is fully explained by classical theory. Because Wien's law relies directly on the logical structure of classical physics, this is cause for concern.

Scene 2. Berlin, late 1890s

Max Planck (1858–1947), a reluctant revolutionary whose "pet subject" was thermodynamics, is drawn to study black body radiation.

In 1878, as a graduate student, he had come across a collection of Clausius's papers, was entranced, and commenced a dissertation that critiqued the existing formulations of the second law. By then, thermodynamics was regarded as virtually complete, and not an exciting or even promising field for a young scientist. But Planck is temperamentally conservative, interested in cementing foundations, and something distresses him about Boltzmann's statistical interpretation of the second law. Laws, he feels, should be absolute—no exceptions, however rare!—and the second law should be as universal as the first, not true by virtue of some statistical sleight-of-hand. In 1895, having moved to Berlin, Planck is goaded by his assistant Zermelo, who produces an argument that the second law can *never* be proven, and, furthermore, that truly irreversible processes are impossible because any mechanical system, however complex, must eventually return to a state near its initial one. Resolving the conflict between the probabilistic, irreversible second law and the ironclad, reversible Newtonian mechanics was "the most important [problem] with which theoretical physics is currently concerned," Planck is moved to say. And black body radiation seemed to hold the key, for the answer might lie in the way the resonators absorb and emit energy. Berlin happens to be the center of studies of black body radiation; Wien is there, as well as several experimenters. Perhaps he could take advantage of their work to show how to use just the theory of electromagnetism plus the laws of thermodynamics to explain the distribution of radiation in equilibrium. Planck begins by revising Boltzmann's work to make it clearer, and reformulating Wien's law in terms of frequency instead of wavelength, seeking to tie all the loose ends of thermodynamics, statistical mechanics, and electromagnetic theory together. And in 1897 Planck gives at the Prussian Academy the first of a series of talks, which end up spanning several years, entitled "On Irreversible Radiation Processes," in which he hopes to solve what he calls "the fundamental task of theoretical physics," the reconciliation of the two laws of thermodynamics. In the very first, he points out the urgent need to look at black body radiation because of a contradiction between the two laws of thermodynamics. The first law of thermodynamics, or "the principle of energy conservation," holds that all effects like friction have to be reduced on a microscopic level to mechanical and reversible processes. But the second law of thermodynamics, or "the principle of the increase of entropy," requires that "all changes in nature proceed in one direction." Reconciling this, he tells his listeners again, is "the fundamental task of theoretical physics." A few talks later, on October 19, 1900, he comes up with an empirical formula that spans both Wien's law at high energies, and classical physics at low energies, as well as that awkward place in between where Wien's law does not quite fit the experimental data. The formula involved "completely arbitrary expressions for the entropy," he says, based on the notion that the resonators could not oscillate at any old frequencies but only at specific ones, related to a number called *h*. Planck is, as usual, very cautious. "[A]s far as I can see at the moment," he says, the work "fits the observational data published up to now as satisfactorily as the best equations put forward for the spectrum." He concludes, "I should therefore be permitted to draw your attention to this new formula which I consider to be the simplest possible."[3] With these words, he introduces—hesitatingly, even reluctantly—the idea of the quantum into physics. That night, one of the experimenters is moved to return to the lab and test Planck's "new formula," confirming it. Planck, excited, sets back to work, and "after a few weeks of the most strenuous labor of my life the darkness lifted and a new, unimagined prospect began to dawn," namely, that he has indeed wrapped thermodynamics, electrodynamics, and classical mechanics in one package, explaining this last piece of the experimental puzzle. It is indeed a neat package, but it has unsuspected implications. Neither he, nor anyone else involved at the time, realizes that in completing the foundations of thermodynamics, they've given birth to an entirely new conception of energy, and come to the threshold of a radically new world.

EPILOGUE

Storm Clouds

In April 1900, Thomson gives a talk at the Royal Institution entitled "Nineteenth Century Clouds over the Dynamical Theory of Heat and Light."[4] The "beauty and clearness" of the theory of heat and light—of thermodynamics and electromagnetism—had been a crowning achievement of nineteenth-century science, he says, but the triumph was "obscured by two clouds."

One cloud was the difficulty of conceiving how the earth moves through the ether. Scientists at the beginning of the nineteenth century had said that the ether must pass through the atoms of solid bodies "like wind blowing through a grove of trees." But Maxwell had shown that the ether must be more like a liquid or elastic solid, exerting force against objects that move through it, implying that the earth's motion with respect to the ether should be detectable. But, Thomson continues, Albert Michelson and Edward Morley had recently carried out an "admirable experiment," flawless in design and execution, that appeared to rule this out. One way out had been independently devised by George FitzGerald and Hendrik Lorentz, who had shown that scientists could "save" the ether if matter flowing through it slightly changed its dimension in the direction of motion—only one hundred-millionth part (the square of the ratio of the earth's velocity around the sun to the velocity of light) would do it! But Thomson found this possibility, though "brilliant," also bizarre. "I am afraid we must still regard Cloud No. 1 as very dense," he concludes.

The second cloud had to do with the "equipartition theorem," as articulated by Maxwell and Boltzmann, according to which molecules store energy by spreading it among all available paths. The theorem provided the explanation for well-known laws regarding the specific heat capacities of solids at high temperatures, but was also in severe contradiction with experimental results involving low temperature solids, gases, and metals. Given the stunning success of thermodynamics, scientists at the time found this discrepancy baffling and scrambled for an explanation. Thomson admits that he had none. He quotes the English physicist Lord Rayleigh, who had daringly said that he was awaiting some new principle that would provide "some escape from the destructive simplicity" of the equipartition theorem. If such an escape should appear, Thomson says in concluding his talk, it would banish Cloud No. 2, which had "obscured the brilliance of the molecular theory of heat and light during the last quarter of the nineteenth century."

Thomson could not know it, but these two nineteenth-century clouds shortly would develop into twentieth-century hurricanes: relativity and quantum mechanics.

Other versions of the drama of the second law of thermodynamics may differ in detail and scope, and number and size of roles. But the drama itself, I claim, is Shakespearean. The cast involves powerful human beings who dedicate themselves body and soul to their work. The action unfolds as these individuals are troubled—sometimes deeply and tragically—by differences between what they find and their expectations, and try to make greater sense of the world by intervening in it. Has any drama ever had such finely drawn and unique characters, or more profoundly reshaped our understanding of ourselves and the world?

INTERLUDE

The Science of Impossibility

Almost every progress in science has been paid for by a sacrifice, for almost every new intellectual achievement previous positions and conceptions had to be given up. Thus, in a way, the increase of knowledge and insight diminishes continually the scientist's claim on "understanding" nature.

—Werner Heisenberg

Many principles of science have the following form: "If you do *this*, what will happen is *that*." Newton's second law, for example, says that the acceleration of a particular mass will be proportional to the force applied to it. Such principles imply that certain effects are practically impossible. A small number of principles, however, belong to a different category. These say, in effect, "*That* cannot happen." Such principles imply that certain effects are physically impossible.

Notorious examples of the latter include the first and second laws of thermodynamics. Other examples include Heisenberg's uncertainty principle and the relativity principles regarding the impossibility of recognizing absolute velocity and the prohibition against faster-than-light travel. Such principles often represent not "new physics" but deductions from other principles. What is different about them is their form. And something about that form—asserting that something is physically impossible—tends to make scientists want to rebel.

The science of impossibility goes by several names. "Forget about it" science is one; "no-way" science is another. Half a century ago, the mathematician and historian of science Sir Edmund Whittaker referred to "Postulates of Impotence," which assert "the impossibility of achieving something, even though there may be an infinite number of ways of trying to achieve it."

"A postulate of impotence," Whittaker wrote, "is not the direct result of an experiment, or of any finite number of experiments; it does not mention any measurement, or any numerical relation or analytical equation; it is the assertion of a conviction, that all attempts to do a certain thing, however made, are bound to fail."

Postulates of impotence thus resemble neither experimental facts that we find in experience, nor mathematical statements that are true by definition prior to all experience. Nevertheless, Whittaker continued, such postulates are fundamental to science. Thermodynamics, he said, may be regarded as a set of deductions from its postulates of impotence: the conservation of energy and of entropy. It may well be possible, he argued, that in the distant future each branch of science will be able to be presented, à la Euclid's *Elements*, as grounded in its appropriate postulate of impotence.

But no-way science is important for another reason: it attracts contrarians. I am not talking about the endless attempts by frauds and naifs to get around the laws of thermodynamics by creating perpetual-motion machines. Rather, I mean serious scientists who find no-way science a challenge to devise loopholes. In seeking these loopholes, they end up clarifying the foundations of the field.

Contrarian science played a role in both the discovery and the interpretation of the uncertainty principle, for instance.

In 1926, Werner Heisenberg was promoting his new matrix mechanics—a purely formal approach to atomic physics—by claiming that physicists had to abandon all hope of observing classical properties such as the position and momentum of atomic electrons, and indeed space and time. Pascual Jordan played the role of contrarian by devising a thought experiment to overcome Heisenberg's claims. Suppose, Jordan said, one could freeze a microscope to absolute zero—*then* one could measure the exact position and momentum of the atom and its constituents. This seems to have inspired Heisenberg to think about the interaction between the observing instrument and the observed situation, putting him on the path that shortly led to his articulation of the uncertainty principle. Jordan, the contrarian, forced Heisenberg to think operationally rather than philosophically, and to clarify the physics of the situation.

Afterward, Einstein famously played the contrarian role—with Niels Bohr as his principal adversary—by trying to devise clever ways of simultaneously determining the position and momentum of a particle. While all his attempts failed, the discussion it provoked did much to help physicists understand the nature and implications of quantum mechanics.

Another famous example of contrarian physics was Maxwell's thought experiment involving a tiny creature who operates a small door in a partition inside a sealed box. By opening and shutting the door, the "demon"—as it was later called—lets all the faster-moving molecules in one side of the partition, violating the second law of thermodynamics by getting heat to flow to that side. The discussion of this thought experiment helped to clarify the then-mysterious concepts of thermodynamics.

Heisenberg is surely overstating when he says that progress in science diminishes the scientist's claim to understand nature: surely the advance of science is more a matter of developing more subtle and complex concepts that replace but also encompass the simpler existing ones. But these more subtle and complex concepts are often produced by those who are dissatisfied by the prospect of having to make the kind of sacrifice Heisenberg mentions.

Dissatisfaction, indeed, is a powerful driving force in science, and it can arise in many ways. The science of impossibility gives rise to a special and rare case of dissatisfaction. This kind of science often collides with our hopes and dreams—of limitless energy, of superluminal travel, of a crisp ontology where things are pinned to specific places at all times. Human beings seem hardwired to have such hopes, and hardwired to balk at the science that dashes them. Small wonder that the science of impossibility makes them dissatisfied. But science benefits in the end.

CHAPTER SIX

"The Most Significant Event of the 19th Century": Maxwell's Equations

$$\nabla \cdot \mathbf{E} = 4\pi\rho$$

$$\nabla \times \mathbf{B} - \frac{1}{c}\frac{\partial \mathbf{E}}{\partial t} = \frac{4\pi}{c}\mathbf{J}$$

$$\nabla \times \mathbf{E} + \frac{1}{c}\frac{\partial \mathbf{B}}{\partial t} = 0$$

$$\nabla \cdot \mathbf{B} = 0$$

DESCRIPTION: A complete characterization of electromagnetism that among other things describes how changing magnetic fields produce electric fields; asserts that there are no magnetic monopoles; describes how electric currents and changing electric fields produce magnetic fields; and describes how electric fields are produced.
DISCOVERER: James Clerk Maxwell
DATE: 1860s; reformulated by Oliver Heaviside in 1884

> From a long view of the history of mankind—seen from, say, ten thousand years from now—there can be little doubt that the most significant event of the 19th century will be judged as Maxwell's discovery of the laws of electrodynamics. The American Civil War will pale into provincial insignificance in comparison with this important scientific event of the same decade.
>
> —Richard Feynman, *The Feynman Lectures on Physics*

Feynman is surely joking again, right? The American Civil War was one of the fiercest conflicts in history. It cost over 600,000 lives, destroyed $5 billion in property, liberated 4 million enslaved people, ended slavery in the U.S., and inflicted economic, political, and social wounds that have never healed. How could this terrible event whose effects are still felt today possibly be overshadowed by some equations, written by a modest Scotsman, who was trying to puzzle out how to describe a few odd effects of little or no practical value?

This time, Feynman was not joking. Maxwell's equations described a new kind of phenomenon—the electromagnetic *field*—that was unanticipated by Newtonian mechanics. These equations characterized this new phenomenon *completely.* They also predicted something novel: the existence of *electromagnetic waves* that could travel through space. And the understanding of electromagnetism that grew out of these equations helped transform it from a curiosity into a *structural foundation of the modern era*, embodied in electronic equipment and in any device based on electromagnetic waves, including radio, radar, television, microwave, and wireless communication. In the process, these equations affected human beings—how they live and interact with each other, themselves, and the world—far more profoundly than any war ever did, or could.

Maxwell

James Clerk Maxwell was born in Edinburgh in 1831, and raised by his parents on a family estate in Glenlair in the Galloway region of southwest Scotland, where he was taught by a private tutor. At the age of ten he was sent back to the city to a school called the Edinburgh Academy for more formal training. There his urbane peers dubbed him "Dafty" for his country attire, strange accent, lack of flair, and childlike questions. But these questions—often a version of "What's the go o' that?"—evidently stemmed from curiosity rather than stupidity. The youth's intellect was further cultivated by William Thomson, the son of a friend of the family who was seven years older than Maxwell, scientifically inclined, and by 1846 already a professor at Glasgow, studying electricity. Maxwell began taking classes at Edinburgh University at the age of sixteen, and Cambridge University 4 years later, winding up at Trinity. After graduating from Cambridge in 1854, the 22-year-old wrote to Thomson that he was interested in studying electricity but was an "electrical freshman."[1] But Maxwell was a quick study, and soon brought himself up to speed.

James Clerk Maxwell (1831–1879)

The field of what was then sometimes called "electrical studies" was in bits and pieces, contributed by a variety of people. The Danish physicist Hans Christian Ørsted (1777–1851) had shown, in 1820, that an electrical current generates magnetism around it. Shortly after, French physicist André-Marie Ampère (1775–1836) wrote an equation, now known as Ampère's law, to characterize this phenomenon mathematically: the total magnetic force around a loop of wire is equivalent to the total current through it. In the 1840s, Maxwell's mentor Thomson (1824–1907) had noticed similarities between the flow of electricity and heat, and wrote equations for electricity to exploit the analogy.

The most extensive investigations had been carried out by British scientist Michael Faraday (1791–1867), who had conducted a long series of experiments before writing *Experimental Researches in Electricity* in 1844. Among other things, Faraday discovered induction—that a moving magnet creates current in a wire, and a changing current creates a current in another wire—and the "Faraday effect"—when polarized light passes through glass in the presence of magnetism its plane of polarization is rotated, meaning that magnetism can affect light.

But Faraday's work was regarded with suspicion by many electrical scientists. They looked on electricity with Newtonian eyes, as caused by a particle- or fluidlike substance that flowed along wires and collected in certain materials, and governed by a force that, like gravity, jumped instantaneously over space to effect action at a distance. To such scientists, what was important to understanding phenomena of electricity and magnetism was the mathematics. Faraday, on the other hand, was convinced that an ether filled all space, and that both electricity and magnetism were caused by strains in this ether and mechanically transmitted by it, probably at some finite speed. As a result, he was convinced that magnetism and electricity affected things like wires and conductors even when these were not moving, creating what he called an electrotonic state. The mathematics was not enough; you needed to understand the activity. As Maxwell wrote later, in contrasting Faraday's views with those of others:

> Faraday, in his mind's eye, saw lines of force traversing all space where the mathematicians saw centers of force attracting at a distance: Faraday saw a medium where they saw nothing but distance: Faraday sought the seat of the phenomena in real actions going on in the medium, they were satisfied that they had found it in a power of action at a distance impressed on the electric fluids.[2]

But Faraday's most serious failing, to his scientific contemporaries, was his lack of mathematical sophistication. Faraday, indeed, was even a little afraid of mathematics and preferred to communicate his ideas in images. He likened strains in the ether, for instance, to "lines of force." He was inspired in part by the fact that, when you sprinkle iron filings over a piece of paper close to a magnet, the filings order themselves up in neat patterns, each filing undergoing induction and turning itself into a little magnet in turn, lining up tip to toe with other filings in smooth curves that depart from one pole of a magnet and return to the other. Faraday came to treat these patterns as observable manifestations of a real something that traversed space. The properties of electricity and magnetism, he felt, derived from how these lines spread, squished, and curved, for which he had only a rudimentary mathematical description. But Faraday's peers felt that, while Faraday's work had much experimental flesh, it lacked mathematical bone.

Maxwell would give mathematical bone to Faraday's experimental flesh. In the process, his impact on electrical studies would be like that of Euler's on mathematics; Maxwell would integrate many areas that seemed independent and even conflicting. His achievement would be so vast and thorough that he would turn what had seemed to be the most independent and thriving of these areas—optics—into the subdivision of a new territory, electromagnetism. But while Euler had reorganized mathematics by fully exploiting the potential of one area—analysis—Maxwell would effect his reorganization, and create this new territory, by a process of analogy. Maxwell's was one of the most brilliant uses of analogy in the history of science, and it helped to bring about one of the most surprising and decisive transformations in civilization's history.

Maxwell's mentor Thomson once said, "I never satisfy myself unless I can make a mechanical model of a thing. If I can make a mechanical model I can understand it. As long as I cannot make a mechanical model all the way through I cannot understand."[3] Maxwell, too, was attracted to the technique. Shortly after graduating from Trinity, he gave a talk to an undergraduate club on the subject—lighthearted in tone, cryptic in argumentative structure, but deeply insightful.[4] Analogies are not about resemblances but about relations, he told the students. Scientists find them valuable because nature is not like a magazine, where you hardly expect that what you find on one page will throw light on the next—but more like a novel, where subjects introduced at the beginning are apt to keep reappearing, in more complex and subtle form, all the way through. Thus exploring the extent to which a strange new phenomenon is like another well-known one, making adjustments where needed, can be a fruitful way to get a grip on the former.

First Step: Mathematical Force

By the time he had given that talk, Maxwell had already begun to use the method to transform the theory of electricity and magnetism. His first step was a paper entitled "On Faraday's Lines of Force," which he read to the Cambridge Philosophical Society in December 1855, when he was 24 years old.[5] The state of electrical science, Part I began, is a mess. We have experimental data for some parts but for others none. Certain pieces have not been mathematicized, while in the pieces that have, the formulas do not all fit together. Anyone studying electricity must mentally store up so much complex and inconsistent information that it is hard to think clearly enough to make a contribution. We must simplify and reduce all the information to grasp it better. I am no experimenter, Maxwell admits, but I shall use physical analogies to develop a mathematics more suited to electrical science. Bear in mind that these are *only* analogies. If we do, we can think more clearly, for we will neither be too distracted by the mathematics on the one hand, nor too stuck on the physical conceptions from which these are borrowed on the other.

Maxwell then mentioned several suitable analogies. One was Faraday's idea that the force exerted by electricity resembles geometrical lines that curve in space. Another was Thomson's idea that electricity flows through space the way heat flows through a fluid: the center of charge is analogous to the source of heat, the effect of electrical attraction or repulsion analogous to heat flow, potential difference analogous to temperature difference, and so forth. A third was the hydrodynamic analogy that an electric charge is like a pump that forces out a stream of an incompressible fluid like water, with speed of the pump like the intensity of the force of the charge, and so forth.

Maxwell continued by assuming Faraday's "vague and unmathematical" idea that an electric field consists of lines of force that spread from one charge to another, and fill all space. Each point on these lines is associated with a direction and intensity. Now suppose, he said in effect, that electricity behaves the way that an incompressible fluid (such as water) does—that is, that lines of force were like tiny tubes carrying the fluid, with the motion resisted by a force proportional to the velocity—but suppose we also correct this picture for the context of electricity by saying that the fluid does not have any inertia. Then a similar mathematical framework for handling fluid flow, which had been developed by Thomson, could be applied to Faraday's conclusions about lines of force. Maxwell uses this picture to put induction and many other of Faraday's various physical ideas—along with Ampère's law—into a set of six laws within a consistent mathematical framework. In Part II of the paper, Faraday handled Faraday's notion of the electrotonic state by developing a variable for it that today is known as the magnetic vector potential (or **A**), couched in a mathematical structure involving differential equations (used to describe properties that change continually over time) that he had learned from Thomson's work. The framework he developed doesn't "*account for* anything" and lacks "even the shadow of a true physical theory," Maxwell admitted; it does not appear to say anything new. But it does provide the "mathematical foundation" of Faraday's researches, which would be a necessary condition for any eventual physical theory.[6]

When Maxwell sent the paper to Faraday, Faraday responded that at first he was "almost frightened" by the application of "such mathematical force" to the subject, but then delighted that the effort succeeded.[7]

Second Step: The Grand Analogy

Maxwell's second step was a paper called "On Physical Lines of Force," written in 1861–62, and it contains one of the greatest uses of analogy in the history of science. Maxwell begins by announcing his intention to "examine magnetic phenomena from a mechanical point of view," and refers to an analogy Thomson had used to understand the Faraday effect: if a magnetic field can shift the plane of polarization of light, Thomson said, it is as if each point on a magnetic line of force were a tiny, spinning "molecular vortex" that passed along some of its spin to any waves of light flowing by.

Maxwell then further develops the image. Let's say a magnetic field consists of such rotating "cells," as he calls them, whose axes are along magnetic lines of force as if threaded on a string; the stronger the field, the more rapidly the cells spin. But Maxwell knows it is mechanically impossible to have cells on neighboring strings spin the same way—clockwise, let's say—for those on one string would rub the wrong way against those in the next. Maxwell rescues the picture by assuming that the space in between is filled with something similar to what engineers call "idle wheels"—smaller beads, in contact with the cells, that rotate counterclockwise, permitting the cells to rotate clockwise. These beads stay in place when the neighboring cells are rotating at the same speed, but changes in the speeds of the vortices cause the beads to move in a line, and they are passed from one cell to another. Thus, Maxwell decides, these beads act much like an electric current.

The model displayed the effects of electromagnetism—the way a changing magnetic field generates an electric current, and an electric current generates a magnetic field—as produced by mechanical motions of a medium. Push-pulls in the ether could produce all the electrical and magnetic effects that Faraday and others had noted.

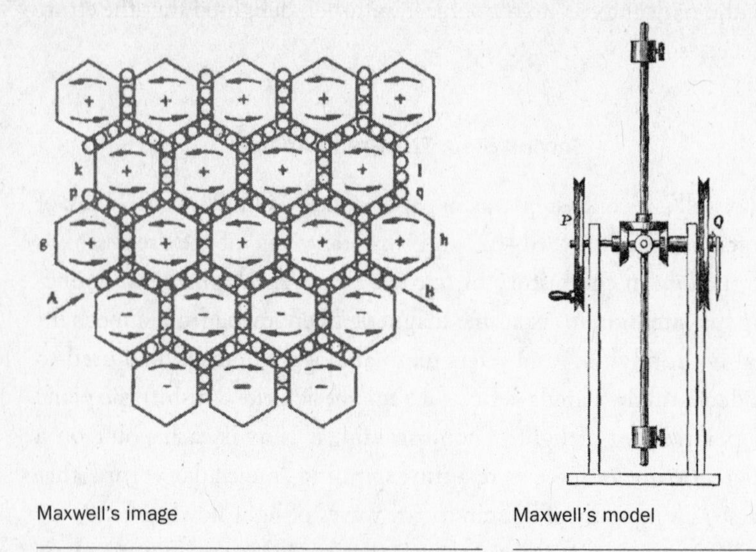

Maxwell's image

Maxwell's model

It even produced a mechanical conception of Faraday's electrotonic state, or what was happening when there was a magnetic field but no electrical current; the electrotonic state was like the impulse of the idle wheels when they turned without moving.

Maxwell wrote up the idea in the spring of 1861, and it was published in installments between March and May. He then left for his usual summer vacation at Glenlair. He was under no illusion that he had created a picture, a representation, of electromagnetism. All he wanted to claim was that this strange model did whatever electrical and magnetic phenomena did, and thus that its mathematics would also work for them. His model was, Maxwell remarked, like an "orrery," or model of the solar system you often see in natural history museums in which the planets are balls placed on rods that mechanically swing about a central ball, the sun. The value of assembling such a model—putting everything you know into it—is that when you finish, and can survey how it works as a whole, you can often see even more than you got from the pieces.

During the vacation, Maxwell realized that he had left something significant out of the model. The cells, he knew, had to have at least a little springiness or elasticity, as do all solid bodies. But this springiness would cause certain effects in his model that he had not accounted for. When the cells pushed the beads but the beads could not move (in an insulating material, say), the cells' elasticity would push the beads a little bit anyway, like rubber balls pushing against an immovable force, until the motion is counterbalanced by forces in the material. If the force were removed, the cells and beads would spring back. Maxwell called this a "displacement of the electricity," whose amount depends on the strength of the electromotive force and the nature of the body. He realized he had to incorporate this into his mathematics, which would also involve introducing a small corrective factor to Ampère's law in the process.

Still more revolutionary: anything elastic can transmit energy from one place to another in the form of waves. Maxwell had shown that the ether—the medium of electrical and magnetic phenomena—must be at least a little elastic. The medium could pass energy in the form of waves from one part to another via leapfrogging electric and magnetic effects operating at right angles to each other—from idle wheels to cells and back to idle wheels again, and on and on, forever. These waves would act the way light does, reflecting, refracting, interfering, and polarizing. Maxwell set out to find the rate that these transverse vibrations travel through the ether, assuming it were passed by purely mechanical forces. The result he calculated, based on the work of Rudolph Kohlrausch and Wilhelm Weber—two German physicists who had measured electrical constants a few years earlier—is 310,740 kilometers, or 193,088 miles, per second. But the velocity of light, as measured by Armand Fizeau a dozen years previously, is 314,858 kilometers, or 195,647 miles, a second, suggestively close. Thus Maxwell wrote, "The velocity of transverse undulations in our hypothetical medium, calculated from the electro-magnetic experiments of MM Kohlrausch and Weber, agree so exactly with the velocity of light calculated from the optical experiments of M. Fizeau, that we can scarcely avoid the inference that *light consists in the transverse undulations of the same medium which is the cause of electric and magnetic phenomena.*"[8]

He published these two revolutionary new features in the model as Part III of his paper in 1862.

Keeping the Baby

Two years later, Maxwell took a third key step in his paper entitled "A Dynamical Theory of the Electromagnetic Field," written late in 1864 and published early in 1865. In it, he cites the earlier mechanical analogy only to abandon it, aiming to present all the results—including the displacement current and the idea that light is an electromagnetic wave—in the form of a set of freestanding equations. "Thus, then, we are led to the conception of a complicated mechanism capable of a vast variety of motion, but at the same time so connected that the motion of one part depends, according to definite relations, on the motion of other parts, these motions being communicated by forces arising from the relative displacement of the connected parts, in virtue of their elasticity. Such a mechanism must be subject to the general laws of Dynamics, and we ought to be able to work out all the consequences of its motion, provided we know the form of the relation between the motions of the parts."[9] Maxwell continued, a few paragraphs later, "In order to bring these results within the power of symbolical calculation, I then express them in the form of the General Equations of the Electromagnetic Field." He then lists twenty equations in eight general categories.[10]

This brought to a close one of the most remarkable uses of analogy in science. His achievement is itself often expressed in terms of a famous analogy—"Maxwell threw out the bathwater and kept the baby"—except that the bathwater begat the baby.

The *Treatise*

In 1873, Maxwell published *A Treatise on Electricity and Magnetism*, his complete presentation of the branch of science that he had developed by his remarkable analogy, and the form in which practically everyone for at least a decade would have to learn it. About a thousand pages long, it was rather difficult and even annoying to digest, for Maxwell made no effort to condense or simplify the work for the reader, aiming to be comprehensive rather than economical. For instance, in the key chapter, entitled "General Equations of the Electromagnetic Field," Maxwell summarizes his work in twelve steps, labeled A to L, each involving an equation or group of equations. "These may be regarded as the principal relations among the quantities we have been considering," he writes. Some could be combined, "but our object is not to obtain compactness in the mathematical formulae." Furthermore, these equations were based on concepts that were extremely difficult to use for those interested in practical applications, most notably **A**, the vector potential, and ψ, the scalar potential.

Maxwell's *Treatise* has also puzzled historians, because in it—and elsewhere—he is silent about how to make and find electromagnetic waves. The idea of electromagnetic waves was the single most thrilling and unexpected feature of Maxwell's entire life's work. His silence over how to make and find such waves seems as perverse as an astronomer whose studies predict the existence of a new planet, yet who does not think to go find a telescope to point at it, or tell someone to go do it. Maxwell's silence is strange enough to demand explanation. Some historians say it is that he was less interested in electromagnetic waves than in light and the ether, others that he did not conceive of any way to produce and detect them, still others that he simply had no time to think on the subject. None of these explanations is really convincing, though it is true that Maxwell's workload had dramatically increased by the time of the *Treatise*. In 1871, he was given charge of supervising the founding of the new Cavendish Laboratory in Cambridge, England, and in 1874, he was handed the task of editing the papers of the laboratory's namesake, Henry Cavendish. Maxwell also became the scientific co-editor of the ninth edition of the *Encyclopaedia Britannica*. These projects left him little time for research.

Maxwell did, however, retain his interest in seeing if the "great ocean of ether," as he called it, could somehow be detected. It is invisible and we know little about it. We do not even know, he wrote in his *Encyclopaedia Britannica* entry on "Ether," if dense bodies like the earth pass through this ocean the way fish pass through water, dragging some small portion of it with them; or the ether might pass though them "as the water of the sea passes through the meshes of a net when it is towed along by a boat." As he wrote beautifully and somewhat anxiously:

> There are no landmarks in space; one portion of space is like every other portion, so that we cannot tell where we are. We are, as it were, on an unruffled sea, without stars, compass, soundings, wind, or tide, and we cannot tell in what direction we are going. We have no log which we can cast out to take a dead reckoning by; we may compute our rate of motion with respect to the neighboring bodies, but we do not know how these bodies may be moving in space.[11]

There is one trick we might play to detect it, he realized, thanks to the fact that a wave flowing through a medium moves at different speeds depending on the speed of the medium. Sound, for instance, always travels at the same speed—about 1,100 feet a second in air—due to the properties of the medium (air molecules) that propagate it. If a wind's blowing, the sound still travels at the same rate in the air, but because the air carries the sound waves along with it, these will seem to be traveling faster or slower than usual from someone on the ground. If a wind's blowing, sound waves thus travel at different rates in different directions.

The same should be true of light. In moving around the sun, the earth might "drag" some small amount of ether with it, but would have some changing velocity with respect to the ether; there would be an ether wind or ether drift. The speed of light would be different in different directions. The difference would be tiny—one part in a hundred million—from the velocity of light in ether at rest. Was this measurable?

On Earth probably not. If experimenters shot beams of light back and forth in different directions, the hundred-millionth-part difference in travel time would be "quite insensible," Maxwell wrote. "The only practicable method is to compare the values of the velocity of light deduced from the observation of the eclipses of Jupiter's satellites when Jupiter is seen from the earth at nearly opposite points of the ecliptic." And so in March 1879, he contacted the director of the Nautical Almanac Office, in Cambridge, England, to ask if any research on this subject had been done. "I am not an astronomer," he wrote with his usual modesty in making the inquiry, but "the only method, so far as I know" of measuring the ether drift would be to make precise measurements of the apparent retardation of eclipses of the satellites of Jupiter.[12]

By this time, Maxwell was showing symptoms of what turned out to be abdominal cancer. That November he died. The career of this prodigious, inventive force of nature—whose quiet pursuits transformed the world, Feynman claimed, more profoundly than the Civil War—died at the age of only forty-eight.

Maxwell left unfinished business—exciting ideas suggested by his work that, for one reason or another, he had not pursued. One was the question of producing and detecting electromagnetic waves; another was measuring the ether drift; a third was revamping his series of equations in a concise way for practical use—which was becoming increasingly important with the expansion of telegraphs. All three of these were carried out in the decade following Maxwell's death.

Heinrich Hertz and the Discovery of Electromagnetic Waves

Heinrich Hertz (1857–1894) was born and raised in Hamburg, and in 1878 began to study in Berlin under Hermann von Helmholtz, who was investigating Maxwell's electrodynamics. Helmholtz tried to entice the bright 22-year-old to compete for a prize to be awarded for the person who solved an experimental problem, devised by Helmholtz himself, that would confirm a certain feature of Maxwell's theory. The youngster declined, afraid the work would absorb several years and not result in a big enough effect to be decisive, and finished his doctoral dissertation instead. In 1885, Hertz moved to Karlsruhe, where he had access to a well-equipped laboratory that he put to use inventively. In 1886, the chance observation that an oscillating current caused sparks to jump across small gaps in a nearby loop of wire set Hertz on a path that led to the publication, in the July 1888 issue of *Annalen der Physik*, of an article entitled "On Electromagnetic Waves in Air and Their Reflection." Hertz was able to measure the wavelength of these electromagnetic waves, and showed they had the properties of other kinds of waves—including the ability to reflect, refract, interfere, and be polarized, and had a finite speed—in stunning confirmation of Maxwell's theory.

Meanwhile, a physics professor at Liverpool in England named Oliver Lodge had noted that oscillating currents created waves in wires. In July 1888, Lodge completed a paper on his results and boarded a train to the Alps for a hiking holiday. En route, he pulled out his reading material—that month's issue of the *Annalen*—to learn of Hertz's work. Lodge was dismayed; he was planning to attend the annual meeting of the British Association for the Advancement of Science that September in Bath and had expected to be celebrated for his discovery, but now realized that Hertz's work would overshadow his. Yet Lodge also found himself thrilled by the elegance of Hertz's experiments, which were much more extensive than his own, for Hertz had detected electromagnetic waves not just in wires but also in air.

The Bath meeting was the first public presentation of Hertz's discovery to the broader scientific community, and the circumstances were rather dramatic.[13] The president of the Mathematics and Physics Section had fallen ill, and his last-minute replacement was Irish physicist George FitzGerald (1851–1901), who had been studying the possibility of producing electromagnetic waves for almost a decade, and who was therefore well-prepared to state the significance of Hertz's work. So while this popular meeting featured a new wax phonograph by Thomas Edison, and a speech on "Social Democracy" by George Bernard Shaw, FitzGerald all but stole the show with the news: electromagnetic force does not work through action at a distance, but by waves traveling through the ether. "The year 1888," FitzGerald announced, "will be ever memorable as the year in which this great question has been experimentally decided by Hertz in Germany." Alerted by FitzGerald's announcement, *Time* magazine called the news "epoch-making." Yet confirmation of Maxwell's ideas also brought to the surface the deep and long-standing dissatisfaction with Maxwell's impractical formulations: FitzGerald spoke of attempts by meeting participants to "murder ψ" and at least revise the vector potential **A**, and the consensus of the gathering was that some conceptual homicide was necessary.

The dramatic news of the creation and detection of electromagnetic waves—implied by Maxwell's work but not discussed by him—also provided a classic illustration of the unexpected productivity of equations themselves. As Hertz once said of Maxwell's equations, "One cannot escape the feeling that these mathematical formulae have an independent existence and an intelligence of their own, that they are wiser than we are, wiser even than their discoverers, that we get more out of them than was originally put into them."[14]

Albert Michelson and the Nonexistence of the Ether

Maxwell's letter about ether drift, sent to the director of the Nautical Almanac Office, was read to the Royal Society at the beginning of January 1880, 2 months after Maxwell's death, and then published in *Nature.* One fascinated reader was American physicist Albert A. Michelson (1852–1931). A graduate of the U.S. Naval Academy in Annapolis, Maryland, who remained there to teach science, Michelson was entranced by the challenge of measuring the speed of light, making attempts in 1878 and 1879, playing hooky from the academy's traditional July 4 celebration to pursue the work. The 1879 measurement, in which he shot a beam of light down a 2000-foot path and back, had an unprecedented precision, earning the 27-year-old a reputation among U.S. scientists and front-page mention in the August 29 edition of *The New York Times*. Michelson's fame, however, did not impress the academy enough to release him from a scheduled sea-duty—but he managed to pull strings and secure a leave of absence, allowing him to travel to Europe at the beginning of 1880 to study physics in Helmholtz's lab. After reading Maxwell's posthumous letter in *Nature* in January 1880, Michelson invented a device, called an interferential refractometer, that used mirrors to split a beam of light by refracting (bending) it, then sent the two beams along two paths at right angles to each other and back. When the two beams were made to interfere, the difference due to their travels through the ether in different directions would be on the order of a fraction of a wavelength—but this tiny difference would be "easily measurable," he wrote to *Nature*.[15] Explaining the planned experiment to his children, Michelson asked them to imagine a race between "two swimmers, one struggling upstream and back, while the other, covering the same distance, just crosses the river and returns." The point, he said, is that "The second swimmer will always win, if there is any current in the river."[16]

A first experiment in 1881 detected no drift, and seemed to have design flaws. Michelson quit active duty, moved to the Case School of Applied Science, in Cleveland, Ohio, and collaborated with Edward Morley (1838–1923), another experimenter, to enlarge and revise the apparatus. This experiment, too, detected no drift, despite an astounding sensitivity of a quarter part per billion. Michelson was baffled and disappointed by the null result, and he and Morley abandoned their plans for further measurements. But other scientists, including George FitzGerald, Dutch physicist Hendrik Lorentz, and French physicist Henri Poincaré, undertook desperate attempts to trust both the Michelson-Morley experiment *and* the existence of the ether, efforts that set the stage for Albert Einstein's discovery of special relativity. In 1907, for his role in the magnificent experiment that made it possible—inspired by Maxwell's letter—Michelson became the tenth person, and the first American citizen, to win the Nobel Prize in Physics.

Oliver Heaviside and "Maxwell's Equations"

The standardization was largely due to Oliver Heaviside (1850–1925), a self-taught electrical engineer, eccentric, and maverick (and discoverer of what was once called the Heaviside layer and now called the ionosphere), who is often called "the last amateur of science."[17] He left home at sixteen, never had a job in a university, and struggled in poverty, supported by relatives, friends, and a government pension. His one and only job was a 4-year stint as a telegraph operator, and he was avidly interested in the practical issue of improving the flow of energy down telegraph wires. He picked up much of contemporary mathematics on his own, using it in novel ways to improve the state of electromagnetic theory; he introduced imaginary numbers into electricity, for instance. When Heaviside came upon Maxwell's *Treatise*, his reaction to it was somewhat the same as Maxwell's own to the then-current state of electrical science: far too complex to be useful to practical folk, for far too many things have to be held in one's head simultaneously. Maxwell's formulation of his theory—founded in the vector potential **A** and the electrostatic potential ψ, a relic of the "action-at-a-distance" perspective—was particularly ill-suited to the increasingly urgent concerns of telegraphy, which involved the flow of electromagnetic energy down specific pathways.

The demands of this practical technology, indeed, did much to advance the science of electromagnetism in the 1880s.[18] Many electromagnetic researchers at the time made clever physical models, involving wheels and connecting bands, to picture to themselves how electrical energy flows from place to place in Maxwell's theory.

Many were frustrated in particular by Maxwell's use of the potentials **A** and ψ.

In 1883, in a series of articles in a magazine called the *Electrician*, Heaviside began to examine how Maxwell's work might be adapted for the practical context of studying the flow of electricity in telegraph wires and circuits. "[I]t was only by changing its form of presentation that I was able to see it clearly," Heaviside wrote later.[19] His amateur, self-taught condition served him well, for he was not inhibited by current mathematical lore nor impressed by prevailing physical perspectives. His outlook was practical; what was important to him was the energy at each point, and calculating how that energy flowed down a path such as a wire. He was prone to expressing that outlook charmingly, in simple and direct terms, as in the following lead sentence from a paragraph in one of his scientific papers: "When energy goes from place to place, it traverses the intermediate space."[20] He then boldly reworked Maxwell's sets of equations in terms of **E** and **H** to represent the electric and magnetic forces at each state, and currents **D** and **B**. The result was a sweeping condensation of Maxwell's work into four equations. These four were pleasingly symmetrical—two electric, two magnetic, and the parallel evident. And they are so thoroughly revamped that they are sometimes called "Heaviside's equations."[21] The equations for free space are the following:

$$\begin{aligned} \operatorname{div} \epsilon \mathbf{E} &= \rho & \operatorname{curl} \mathbf{H} &= k\mathbf{E} + \epsilon\dot{\mathbf{E}} \\ \operatorname{div} \mu \mathbf{H} &= 0 & -\operatorname{curl} \mathbf{E} &= \mu\dot{\mathbf{H}} \end{aligned}$$

and in their more complicated form in the presence of electric charges

$$\begin{aligned} \operatorname{div} \epsilon \mathbf{E} &= \rho & \operatorname{curl}(\mathbf{H} - \mathbf{h}_0 - \mathbf{h}) &= k\mathbf{E} + \epsilon\dot{\mathbf{E}} + \mathbf{u}\rho \\ \operatorname{div} \mu \mathbf{H} &= \sigma & -\operatorname{curl}(\mathbf{E} - \mathbf{e}_0 - \mathbf{e}) &= g\mathbf{H} + \mu\dot{\mathbf{H}} + \mathbf{u}\sigma \end{aligned}$$

Heaviside himself modestly referred to his four equations as "Maxwell Redressed,"[22] though he did promote them, enthusiastically and polemically, as superior to Maxwell's own equations and to other revisions thereof. Shortly after the 1888 Bath meeting, for instance, he published a brief note savagely attacking the continued use in propagation equations of the electric potential ψ and the vector potential **A** as "metaphysical" (a term of opprobrium for scientists) and as "a mathematical fiction."[23] What we measure, after all, are the electric force **E** and the magnetic force **H**, not potentials. These give us real information about the state of the field; these are what propagate when current flows. Keeping ψ and **A** results in "an almost impenetrable fog of potentials" and even inconsistencies, and Heaviside, recalling the Bath conference, advocated their "murder." Maxwell's theory works just fine, he concluded, "provided that we regard **E** and **H** as the variables."

Heaviside's version of Maxwell's equations were quickly and gratefully adopted by prominent electromagnetic researchers, including Hertz, and the entire scientific community converted by the 1890s. The equations have remained virtually the same ever since; the version at the beginning of the chapter is taken from the standard textbook *Classical Electrodynamics* by J. D. Jackson.

Fittingly, Heaviside's achievement in revising Maxwell's was once captured in an analogy. Reviewing Heaviside's work, FitzGerald compared Maxwell to a general who conquered a new territory but had not the time to find the best roads or make a systematic map. "This has been reserved for Oliver Heaviside to do," he wrote. "Maxwell's treatise is cumbered with the *débris* of his brilliant lines of assault, of his entrenched camps, of his battles. Oliver Heaviside has cleared those away, has opened up a direct route, has made a broad road, and has explored a considerable tract of country."[24]

But Maxwell was a strange kind of general who worked in a strange kind of terrain. The territory he conquered was so potent and extensive that his work, as Feynman noted, would have a far greater impact on human nature than that of any group of generals.

Overcoming Anosognosia; or Restoring the Vitality of the Humanities

Simon Schama's book *History of Britain*, at 1,500 pages, is a solid history of that country, and has been made the basis for a multipart documentary film. Yet the book contains no mention of James Clerk Maxwell, nor any mention of the role that this scientist played in laying the foundation for electrification, light, heat, communication, and the electronics revolution of the twentieth century, in Britain or elsewhere. Schama's book omits any references to the contributions made by British scientists and engineers to transforming Britain and the world.

The neglect of science, indeed, is common in history books—most disturbingly, even in books that profess to care about the masses, and oppressed and underprivileged peoples. Since it was first published in 1980, for instance, *A People's History of the United States*, by Howard Zinn, has sold over a million copies and become one of the most influential works of history in the U.S. A popular textbook in schools and colleges, it claims to focus on "hidden episodes of the past when, even if in brief flashes, people showed their ability to resist, to join together, occasionally to win."

However, Zinn's book makes no mention of people resisting, joining together, and winning when it comes to science. It says nothing, for instance, of the struggles to reduce childhood mortality, to increase life expectancy, or to develop systems of mass transportation. There is no mention of Norman Borlaug, who won the 1970 Nobel Peace Prize for leading the "green revolution," and who helped end hunger for millions of people. Another no-show is the microbiologist Maurice Hilleman, whose vaccines saved more lives than were lost in all the wars to which Zinn devotes chapters.

Mass electrification fails to feature in Zinn's book, although the unit costs of electricity are discussed in the context of a program to give "enough help to the lower classes" to prevent them rebelling. Steam power is not covered, nor is the internal-combustion engine, although railroads are discussed in relation to racial segregation, unions, strikes, and methods of exploiting American Indians.

Zinn, in short, considers scientific changes inconsequential to "the people." History, for him, is a grand pageant of ideologies; if science is at all significant in that pageant it is perhaps only in forging the weapons that the ideological partisans use to beat up each other.

The omission does not necessarily make the book defective as history. As Zinn notes, historians cannot avoid selecting and emphasizing some facts rather than others, although they have a duty to avoid promoting ideological interests, knowingly or not. But Zinn's omissions do make the book defective as an account of "the people." The conquest of dreaded and once-common epidemic diseases, such as polio and encephalitis, have fundamentally affected how all of us view life and death. Developments in astronomy and the discovery of evolution have affected our sense of time and space, and our place in nature. These events all took place within the timeframe of Zinn's book. Although some of these developments were pioneered by non-Americans, they profoundly altered how human beings seek answers to the questions of what we know, should do, and can hope for.

Schama and Zinn are not the only ones to ignore the impact of science. Many authors of contemporary fiction fill their books with characters who are nothing more than superannuated children, seemingly unaffected by technological training and devices. Some writers—like Jonathan Franzen, Ian McEwan, Neal Stephenson, and David Foster Wallace—do present protagonists who are interested in and influenced by their technological surroundings. But these writers can be severely criticized by reviewers for their efforts.

Commenting on McEwan's *Saturday*, for instance, John Banville roasted the author for being "wearingly insistent on displaying his technical knowledge" and complains of "big words in this book." The book indeed has some big words. However, the training that turns people into technically literate professionals not only accustoms them to using big words, but also affects how they speak and act. Technically competent people often delight in their technical competence, and wield this competence when interacting with the world. This is precisely what McEwan so ably captures.

Dismissing the effect of science on modern life has nothing to do with the "two cultures." Rather, it shows a blind spot in the work of some writers and scholars whose duty it is to become aware of the world around them. It is more serious than amnesia. We can name the condition with one of the "big words" that McEwan's protagonist uses in *Saturday*. It is "anosognosia"—a medical term (derived from a combination of the Greek words *agnosia*, or "without knowledge," and *nosos*, or disease) that means a lack of awareness of one's own diseased condition; that is, not knowing that one is diseased.

What are the causes of anosognosia? I count four contributing factors.

One is drama: scientific and technological change tends to lack the exciting settings of other historical turning points. It is not generally heralded by bloody battlefields or by clashes of titanic personalities, and unfolds in a way that makes it difficult to dramatize differences. A second is the hope among even so-called enlightened and progressive scholars that we can reinvent ourselves and remake the world, Marxian-style, achieving liberation at a revolutionary stroke; admitting dependence on science and technology serves to dampen such hopes. A third is fear of specialized knowledge, knowledge that one might take extra training to acquire.

Finally, and most importantly, scholars in the humanities often see themselves as having a critical function—they see themselves as asking the important questions that help humanity navigate the world's dangers. But if the fate of "the people" is as tied up with science and technology as it is with ideologies—with who is exploiting whom—this leading role is blunted, or at least shared. For those who identify the humanities with such a critical function, this might even seem threatening. Far safer for its practitioners to circle the wagons, dwelling on what is distinctive about the humanities rather than what is possible! This is what makes so many humanities programs both defendable and lifeless. Moreover, such wagon circling is self-interest in disguise; thus, an ideology—a belief structure lacking empirical support—itself.

Overcoming anosognosia requires admitting that a truer picture of humanity may be less dramatic than we hope, curbing our fascination with shortcuts to liberation, and accepting that humanity's important questions are addressed by a variety of disciplines. This will strengthen, not threaten, the humanities. For only when the humanities couple their inquiries into human dimensions and possibilities with an awareness of what science has disclosed of the dimensions and possibilities of the world will the humanities most effectively be able to provide answers to the questions of what we know, should do, and can hope for.

CHAPTER SEVEN

Celebrity Equation

$$E = mc^2$$

DESCRIPTION: Energy and mass can be converted into one another, with the amount of energy being equal to the mass multipled by the speed of light squared.
DISCOVERER: Albert Einstein
DATE: 1905

A while ago I was reading an interview with the actress Cameron Diaz in a movie magazine. At the end the interviewer asked her if there was anything she wanted to know, and she said she'd like to know what $E = mc^2$ really means. They both laughed, then Diaz mumbled that she'd meant it, and then the interview ended.

—David Bodanis, *$E = mc^2$: A Biography of the World's Most Famous Equation*

$E = mc^2$ is the most famous equation of all time. It has made the cover of *Time* magazine. It has been the subject of a "biography" that treated the equation as though it were a person. It is the title of a play by Hallie Flanagan, the woman who headed the Federal Theatre Project during the Depression. The Dalai Lama calls it "the only scientific equation I know."[1] Poems and pop songs have been written about it; those of a certain age may remember the hit single "Einstein A Go-Go," by 1980s electronic pop band Landscape, the lyrics of which went "You'd better watch out, you'd better beware, coz Albert says that E equals mc squared." More recently, singer Mariah Carey put out an album entitled, $E = MC^2$, with the right-hand term alluding to her initials. During the so-called science wars of the 1990s, debate raged over the French feminist philosopher Luce Irigaray's assertion that $E = mc^2$ is a "sexed equation" because it privileges the speed of light.[2] The equation has turned up on postage stamps of various lands, in movies (*School of Rock*), popular fiction with scientific pretensions (Dan Brown's *Angels & Demons*), and numerous cartoons and video games.

The physicist Stephen Hawking was once warned not to include any equations in his writings for a general audience because, or so he was informed, every equation would halve the number of readers. As a result, he was determined not to use any equations in his book, *A Brief History of Time*. But $E = mc^2$ appears in the book, and in multiple places. This did not dent sales, and it went on to become one of the best-selling science books for a general audience of all time.

All this might make us wonder whether $E = mc^2$ is not a real equation at all but rather a celebrity. A celebrity is someone everybody knows of, but not about. Similarly, everybody recognizes this equation, and is sure that it is important, but it's never clear exactly why. We know plenty of gossip about it but still always feel we are seeing it from the outside. We wonder how much work it really does. The status of $E = mc^2$, like that of a celebrity, seems manufactured by some mysterious social process.

Yet, in the end, celebrities are just human beings, and $E = mc^2$ is just another equation. Like other equations, it sprang from dissatisfaction with the way things were fitting together, its first appearance was different from the form in which we know it today, it reorchestrated the way human beings looked at the world, and it had unexpected consequences.

How, then, did this equation get to be a celebrity?

The Collision Between Newton and Maxwell

Equations can be born from several different kinds of dissatisfactions. Some spring from a scientist's sense that a confusing heap of experimental data can be better organized. Others arise from the feeling that a theory is too complicated and can probably be simplified, or that its parts are not fitting together properly. Still other dissatisfactions arise from mismatches between a theory's predictions and experimental results.

The equation $E = mc^2$ resulted from a special and rare case of dissatisfaction felt by many physicists at the end of the nineteenth century and the beginning of the twentieth. The dissatisfaction was created by a troubling experimental result that highlighted an inconsistency between two great, comprehensive, and venerable scientific systems: Newton's and Maxwell's. More exactly, the result highlighted an inconsistency between two principles—the principle of the relativity of motion, and the principle of the constancy of the speed of light—each basic to one system.

The inconsistency involved an idea called *invariance.* In its loosest sense, invariance simply means that something can appear in two different ways but nonetheless be the same thing. Two people standing in different parts of a room, for instance, can see a chair as to the right or to the left of the television—but once we take the difference in positions into account, it becomes clear not only that they are seeing the same chair, but also how and why that chair appears differently to each. If we cannot account for the difference in appearance, then one or both of these people is hallucinating or seeing some sort of illusion. Real things, we might say, are *supposed* to look different from different perspectives. Reality therefore necessarily involves a difference between how something appears to us, and what it really is. We can put this point another way, in terms of the difference between local effects and global properties. When I see an object I see only one profile of it—one that changes if I move, if the light changes, and so forth. As I change my position, so does this "local" effect. Yet all the time I am seeing the same object. Invariance therefore involves understanding unity as it shows itself in changing appearances. Philosophers call this the noetic-noematic correlation; physicists call it invariance under transformation, or covariance. Covariance is simply part of the definition of objectivity; to say that something is a real part of the world is to say that it looks different from different perspectives, though the descriptions flow together in an orderly way when described by the right set of transformations.

Newton's mechanics supposed the existence of an absolute time and an absolute space as the arena in or stage on which events happen, and that there is no privileged time or space on that stage. This is called translation invariance, and all it means is that if we move about in time or space the laws are still the same. But Newtonian mechanics implies a further kind of invariance, that according to a "principle of relativity of motion," there is no privileged movement, whether in motion or at rest. The laws of physics are the same for anyone moving at constant speed regardless of direction, no matter how fast or slow. This is a familiar experience. So long as a train, say, is traveling smoothly and doesn't jostle, anything we do—drink a glass of water, play cards or handball, dance—happens exactly the same as if the train were at rest in the station. The water stays in the glass and doesn't slosh, the ball bounces at the equivalent point on the floor, and the dancer confidently executes the same gestures and winds up in the same spot as if the train were still. We could perform no experiment to tell how fast the train was moving, or even whether it was moving. Even people on a second train at rest in the station down the track would see the same laws of physics in play on ours, once the difference in speed between the trains was taken into account. And that difference in speed is a matter of simple addition and subtraction.

Scientists would call such a train, from whose position we describe events, a *reference frame*, and one moving at uniform speed an *inertial reference frame*. They call *transformations* the equations used to change the mathematical description of an event—its x, y, and z of position and its time t—from one reference frame into another. They call the equations that connect the properties of a description in one inertial reference frame with those of another *Galilean transformations*, for these express a principle of relativity of motion already present before Newton in Galileo's mechanics, in the latter's thought experiments involving dropping cannonballs from the masts of sailing ships. The Galilean transformations are quite simple. On board the moving train, for instance, the only thing about events that changes is their distance down the track (let's make that the x-axis). Any x position on that train, call it x', differs from the x position for an observer on the ground by the distance the train has traveled in a time t: $x' = x - vt$. All the other coordinates—y and z—remain the same, and things continue to happen at the same time t.

A physicist's definition of reality and objectivity depends on Galilean transformations. A "real" thing or event is one with the same physical description in different inertial frameworks, once you use the appropriate transformations to take the differences in speeds and directions into account. The notion of reality *requires* drawing a difference between how something appears to us, and how we describe it; the variability to observation is built into the objectivity of the object that I see. In developing the notion of transformations, scientists were merely elaborating the conditions of objectivity—of what is the same regardless of which inertial frame it is seen from.

Thus the principle of the relativity of motion was at the core of Newtonian mechanics. But according to a "principle of the constancy of the velocity of light" central to Maxwellian mechanics, light introduces a new element into this neat picture. Light acts more like sound. Sound always travels at the same speed (about

1,100 feet a second in air), regardless of how fast its source travels. The reason has to do with the properties of the medium (air molecules, say) that propagate the sound waves that make it impossible to push sound waves any faster than a certain speed. According to Maxwell's equations, light also always travels at the same speed (about 186,000 miles a second) regardless of the speed of the source. Physicists assumed this stemmed from the fact that light moves in a medium called ether, whose properties governed how fast light could travel. If so, this principle implied that there *was* a favored inertial reference frame in the "stage" of absolute space and time, provided by the ether. In moving around the sun, the earth moves through the ether, and while it might "drag" some small amount with it, its speed with respect to the ether could be detected by measuring the speed of light in different directions. For the ether moves the light, by an amount that involves the Pythagorean theorem.

Imagine a speedboat heading toward the opposite bank of a 400-yard river, whose flow carries it 300 yards downstream in the process. The water, of course, moves the boat. The speedboat winds up moving at an angle along the hypotenuse of a right triangle: 400 yards across and 300 yards downstream; it has traveled (because we've conveniently based the example on a Pythagorean triplet) a total of 500 yards in crossing the river. To end up directly across the river, it would have to point itself upstream while traveling by the same angle, and will in effect travel a longer distance (the hypotenuse of the above triangle) in crossing the bank directly. For the same reason—so the argument ran—the ether would move the light traveling in it, and the light would travel at a different speed when crossing the ether's direction of motion.

In 1881 and 1887, two American physicists, Albert Michelson and Edward Morley, carried out an extremely sensitive experiment to detect what scientists were calling the "ether drift." Their instrument consisted of two "arms," one pointing in the presumed direction of the ether's motion and the other perpendicular to it, along which beams of light would travel back and forth. Mirrors were mounted on a bed of mercury, then rotated 90 degrees so that the light would be traveling in a different direction with respect to the ether's motion. By bringing the beams together to create an interference pattern, Michelson and Morley would be able to detect any slight difference in their velocities. But the experiment failed to detect any such difference.

Physicists were baffled. Something was clearly wrong either with Newton's or Maxwell's equations.

At first they assumed the problem lay with Maxwell. He was the Johnny-come-lately. Maxwell's equations had been around for only a few decades, while Newton's laws had been around for two hundred years and successfully accounted for everything except a few disagreeable but minor discrepancies that there was little reason to think would not prove to be due to some experimental error or overlooked effect. Some of the brightest scientists of the day attempted to modify Maxwell's equations to fit the Galileo transformations.[3] But these equations proved remarkably resistant. They were embedded in an elaborate network of interrelated concepts, and any change in one rippled through the others with undesirable results.

As the nineteenth century drew to a close, many physicists interested in electrodynamics felt a deep dissatisfaction. There *had* to be an explanation—the constancy of the speed of light regardless of direction of motion had to be reconcilable with Maxwell and Newton—but none could be found. "The most incomprehensible thing about the world is that it is comprehensible," Einstein once declared. The unstated corollary, is that, to a scientist, the most frustrating thing about the world is not being able to comprehend it.

Acts of Desperation

Dissatisfaction led to desperation. In 1889, Irish physicist George FitzGerald wrote a short, one-paragraph article—a mere five sentences, no equations—stating that "almost the only hypothesis" that can reconcile the Michelson-Morley experiment with Maxwell and

Newton "is that the length of material bodies changes, according as they are moving through the ether or against it by an amount depending on the square of the ratio of their velocities to that of light."[4] Suppose, FitzGerald thought, the arm of Michelson and Morley's instrument pointing in the direction of motion shrank due to the impact of the ether on its molecules. If it shrank by just the right amount, it would "measure" the light beam as going up and down in the direction of the ether at the same speed as it measured the light traversing the perpendicular arm. Still, the idea—objects shrink in size when moving at high speeds?—seemed too bizarre to take seriously.

Another desperate soul was Dutch theorist Hendrik Lorentz, who wrote to his friend Lord Rayleigh in 1892 about the predicament created by the Michelson-Morley experiment, "I am utterly at a loss to clear away this contradiction."[5] That year, he independently proposed the same idea that FitzGerald had, writing that "I can think of only one idea" to explain the experiment, namely, that the ether causes some contraction effect in the length of a solid body. When he learned of FitzGerald's idea and contacted him, FitzGerald was overjoyed to learn of a fellow champion of contraction, writing back that he had been "laughed at" for his ideas.[6] Lorentz then went on to work out in detail the set of transformations that would have to occur for this contraction to work. Lorentz found that time, too, would be affected. For while FitzGerald was only trying to save the Michelson-Morley experiment, which determined the light in two directions to be traveling at the same speed, Lorentz, more ambitiously, wanted to make sure that the speed of light stayed constant, and appeared the same to moving and nonmoving observers. To accomplish this, clocks would have to tick slower. He then produced a set of formulas now known as the Lorentz transformations, which gave compensations in length and time between stationary and moving systems that would preserve the possibility of the light moving at a constant speed in the ether as detected by the Michelson-Morley experiment, and thus the agreement between Maxwell and Newton. The compensation factor for both space and time was $\sqrt{1 - v^2/c^2}$.[7] Notice that when there is no relative motion (and v is 0), there is no correction. At low speeds, the correction is so small it would not be noticed. But the closer the object approached the speed of light, the larger the corrective factor grew—the more the object shrank in the direction of motion, and the slower clocks ticked. Meanwhile, however, most scientists continued to regard the idea as too strange to take seriously. But the fact that it was said at all shows the lengths to which scientists were prepared to go to save the ether.

It seemed, one scientist would say later, that "all the forces of Nature had entered on a conspiracy" with the goal of "preventing us from measuring or even detecting our motion through the ether."[8]

The consternation mounted. Great scientists began reaching for fantastic ideas. In 1898, French mathematician Henri Poincaré toyed with the idea of giving up absolute time in favor of "local time," and soon tried to use it to explain the puzzle regarding the speed of light in ether. In a public lecture at the 1904 World's Fair in St. Louis, Poincaré remarked almost whimsically, "Perhaps we should construct a whole new mechanics, of which we only succeed in catching a glimpse . . . in which the velocity of light would become an impassible limit."[9]

Thus the problem that Lord Kelvin had called "Cloud No. 1" obscuring the "beauty and clearness" of nineteenth-century dynamical theory was only getting more and more difficult. The year after Poincaré spoke, in 1905, all the fantastic ideas that had been cited to try to banish it—the contraction of space and time at high velocities, the nonexistence of absolute space and time, and the speed of light as an absolute upper limit—were shown, in one form or another, to be true.

Enter Einstein

How had Einstein, at that time still a patent clerk, come to take up this particular problem? The same way everyone else did: dissatisfaction.

Years later, Einstein wrote to a friend that, while only about 5 or 6 weeks elapsed between his conception of the idea of the special theory of relativity and a finished paper about it, "the arguments and building blocks were being prepared over a period of years."[10]

The earliest argument emerged in late 1895 or early 1896, when Einstein was sixteen. It came in the form of what he would call a "childlike thought-experiment." (The adjective "childlike," in the sense of pure and direct, was often applied to Einstein.) What, the youth asked himself, would happen if he were traveling at the speed of light, and looked over at a light beam riding next to him?[11] Newton said it could happen, Maxwell said it couldn't.

This simple puzzle—can you or can't you catch up to a light wave—had to have an answer, but none could be fashioned from the existing tools of physics. The puzzle focused the dissatisfaction of young Einstein, providing him with the mixture of bewilderment and curiosity needed for him to start fashioning the arguments and building blocks.

Einstein brooded for years. "I must confess," he told a friend later, "that at the very beginning, when the Special Theory of Relativity began to germinate in me, I was visited by all sorts of nervous conflicts. When young, I used to go away for weeks in a state of confusion, as one who at that time had yet to overcome the stage of stupefaction in his first encounter with such questions."[12] One day in 1905 he went to visit his close friend and patent office colleague Michele Besso, poured out the details of his "battle" with the problem, and departed. But in the process of laying out the problem, Einstein found the solution. The next day, he dropped in on Besso again and greeted him with the words, "Thank you. I've completely solved the problem."[13]

The result was "On the Electrodynamics of Moving Bodies," one of the most famous and momentous scientific papers ever written, sent to the journal *Annalen der Physik* in June 1905. Despite the angst behind the genesis of the paper, it follows a simple yet powerful logic—"a deep, almost childlike freshness of approach"[14]—that is relatively easy to understand.

"It is well known," Einstein begins, "that Maxwell's electrodynamics—as usually understood at present—when applied to moving bodies, leads to asymmetries [eccentric results] that do not seem to attach to the phenomena [that is, they seem to be an artifact of our theories rather than a part of the world]."[15] He gives examples, and says that these, "and the failure of attempts to detect a motion of the earth relative to the 'light medium,' " lead to the postulate that there is no such thing as "absolute rest." He calls this conjecture "the principle of relativity," and says that he will combine it with the postulate that in empty space "light is always propagated with a definite velocity V which is independent of the state of motion of the emitting body."

Thus Einstein framed his paper around the logical requirement of reconciling the two key principles: relativity and the constancy of the speed of light. These are "seemingly incompatible," Einstein says. Only seemingly. For he develops a reconciliation in the rest of the paper, claiming that, based on logic alone, he can produce a "simple and consistent electrodynamics of moving bodies" with no need for the supposition of an ether or for an absolute rest frame. What would it take for observers on two different inertial frames to see light travel at the same speed? Einstein determines that it would require the same contraction factor for length in the direction of motion and time that Lorentz proposed. But while Lorentz had based his work (as had FitzGerald) on the assumption that the ether existed and that the contraction was real (due to the effect of ether on molecular forces), Einstein based his only on the assumption of the validity of the principles of relativity and of the constancy of light. That is, while Lorentz and FitzGerald got their results by trying to save the ether, Einstein arrived at the same result by getting rid of it. As scientists said at the time, "There is no conspiracy of concealment, because there is nothing to conceal." Or as Feynman liked to say, a universal conspiracy is a law of nature.

Einstein refers to the contraction factor as "β" in his paper. Its deduction is most easily and frequently presented as a Pythagorean

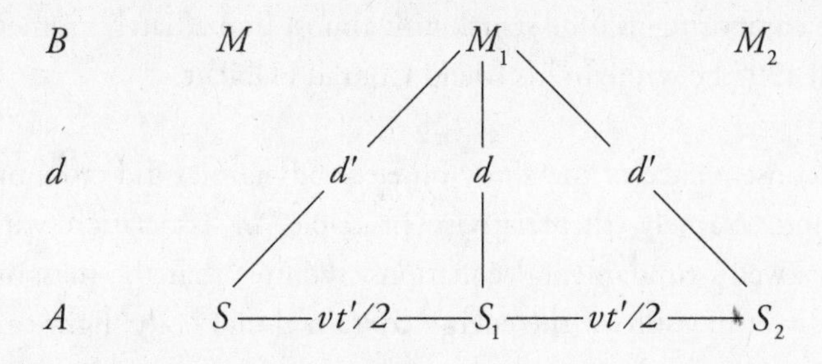

problem. Suppose two inertial reference frames, *A* and *B*, are moving at velocity v with respect to each other. In *A*, a beam of light is sent from a source, perpendicularly to the direction of motion, to bounce back off a mirror at a distance *d* away from the source. From the point of view of someone on *A*, the light simply travels a distance $2d$. But to someone on *B*, for whom *A*—source, mirror, and all—is gliding past at a velocity v, the light travels a longer path; we'll call it $2d'$. Half of this path, d', is the hypotenuse of a right triangle whose other sides are d and $vt'/2$. Thus $(d')^2 = (d)^2 + (vt'/2)^2$. Yet according to the second principle, the light has the same speed, c, covering the same distance in the same time, seen from *B* as it does seen from *A*. That is, V (the symbol Einstein is using for the speed of light) is equal to $2d/t$ in *A and* to $2d'/t'$ in B. How can that happen? Only if the distance and time of objects in *A* are shorter in *A* as seen from *B*. By how much? By just the amount that d is shorter than d'; that is, d/d' or t/t', or the contraction factor β. If $V = 2d/t$, then $d = Vt/2$; and if $V = 2d'/t'$, then $d' = Vt'/2$. Substituting in the Pythagorean equation gives us β (or the contraction factor t/t' we are seeking) $= \sqrt{1 - v^2/c^2}$.

This, the seminal paper of what would become known as the "special theory of relativity" (a usage that Einstein began in 1915, to distinguish it from his then-new "general theory of relativity"), was published in September 26, 1905. It introduced some radical changes in notions of space and time. A paper with such fundamental implications, however—especially one put together in such a short time by someone working feverishly on so many things—was bound to have more consequences than its author could foresee while composing it. One struck him almost immediately. Sometime in fall 1905 he wrote to his friend Conrad Habicht,

> A consequence of the study on electrodynamics did cross my mind. Namely, the relativity principle, in association with Maxwell's fundamental equations, requires that the mass be a direct measure of the energy contained in a body; light carries mass with it. A noticeable reduction of mass would have to take place in the case of radium. The consideration is amusing and seductive; but for all I know, God Almighty might be laughing at the whole matter and might have been leading me around by the nose.[16]

Being led "around by the nose"—reminiscent of how Meno's slave must have felt, learning something which appears to be true, yet which also must be further explored.

Einstein mailed a three-page paper outlining this consequence, entitled "Does the Inertia of a Body Depend Upon its Energy Content?" to the *Annalen* the day after his relativity paper appeared, and it was published later that year. As historian of science John Rigden, among others, has pointed out, this paper does not break new ground, and simply draws a consequence that was logically implicit in the previous paper, and easily could have been its final section. If it had, Rigden says, "it would have made a spectacular conclusion."[17]

Einstein opens the "Energy Content" paper in a disarmingly modest key, "The results of an electrodynamic investigation published by me recently in this journal lead to a very interesting conclusion." He reaches the conclusion via the following example. Suppose an object (an atom, say) of mass m at rest in reference frame A emits two beams of light—thus, it expends energy—in opposite directions. Let's say the total amount of energy lost is L (as in the previous paper, Einstein uses the now-unfamiliar notation of L for energy and V for the speed of light), so each light beam carries away the energy $L/2$.

An observer on *A* sees the object as having no net change in kinetic energy; the atom is standing still, has shed some of the energy it had in an excited state, and continues to have the same mass that it was originally stamped with. But an observer in *B*, for whom *A* is moving, sees something different. The forward-moving light beam has more momentum than the backward one, meaning that the atom has had a net change—a decrease—in kinetic energy. This can happen only if the atom's velocity or its mass decreases. But its velocity is the same; in the rest frame, there is no recoil. The only other possibility is that, from the perspective of the frame in which the atom is moving, the mass has decreased. The atom has not gained any mass from the point of view of its rest frame; its "inertial mass" is the same. But its mass from the point of view of the laboratory, which views it as a moving object, changes. By how much? Applying the tools of the previous paper, Einstein finds that the conversion factor, once more, is β.

Einstein continues—again, using the unfamiliar notation of *L* for energy and *V* for the speed of light—as follows:

> If a body releases the energy L in the form of radiation, its mass decreases by L/V^2. Since obviously here it is inessential that the energy withdrawn from the body happens to turn into energy of radiation rather than into some other kind of energy we are led to the more general conclusion: The mass of a body is a measure of its energy content.[18]

This is the first appearance in print of the idea eventually to become famous as $E = mc^2$. It is not presented explicitly in the form of an equation, and it is not in its familiar symbols. However, the startling, even revolutionary mass-energy concept is fully articulated. The concept transformed some of the most fundamental notions of how the universe is assembled. It put together two things long thought to be utterly different: energy, whose conservation principle was a crowning achievement of nineteenth-century physics, and mass, whose conservation principle was a crowning achievement of eighteenth-century science.[19] They can change into each other.

It also revolutionized the requirements for objectivity. On the Newtonian stage, energy and mass remained the same when observed from different inertial frames; on Einstein's, they remain virtually the same at low speeds, but changes take place the closer the speed of the frames get to that of light. What is objective—really out there—is what changes in length and clock time by this amount when witnessed from another, sufficiently fast, inertial frame.[20]

Over the next several years, Einstein referred to this result several times, though again in the form of descriptions or in his original symbols, and not yet in his now-famous version. In a footnote to a 1906 paper, for instance, Einstein wrote that "the principle of the constancy of mass is a special case of the energy principle."[21] Early in 1907, in another *Annalen* article, he refers to energy as ϵ, mass as the Greek letter μ, and the speed of light as V, and he uses the equation

$$\epsilon = \mu V^2 \frac{1}{\sqrt{1-(v/V)^2}}$$

Here the famous equation—energy is equal to the mass times the speed of light squared—appears with the corrective factor β, which takes into account the effect when the body is in motion. That is, let's take a piece of matter such as an electron. At rest, every electron has the same mass; it's as if nature stamped that mass on each electron when it was created. Whenever that electron is weighed in its own reference frame, it always has that mass. Now suppose we look at it from another reference frame, in which the electron is moving. If $E = mc^2$, and c is a constant, then m and e have to vary in exactly the same way as the energy increases. The electron's inertial mass—its mass from its own rest frame—does not change. But its mass as you measure it in the laboratory, where you see it as moving, changes. And β, the compensation factor, is the transformation that tells you what to multiply by to get the rest mass. Leaving out that compensating factor gives you what he calls in a footnote the "simplifying stipulation $\mu V^2 = \epsilon_0$."[22]

Later that year, Einstein changed his symbols to use c rather than V for the speed of light. The theory of relativity contains a result of "extraordinary theoretical importance," he says; that "the inertial mass and the energy of a physical system appear in it as things of the same kind. With respect to inertia, a mass μ is equivalent to an energy content of magnitude μc^2."[23] Over the next few years, Einstein worked out more fully the mass-energy principle and its implications. And in a manuscript on relativity theory of 1912, at the beginning of a discussion of the subject, he writes the above formula using m in place of μ, and a script L (as in the first version) in place of ϵ then crosses it out and writes E. From then on, he sticks with E and c, and we now have the familiar equation plus the corrective factor where q (sometimes written as v) stands for the velocity:[24]

$$E = \frac{mc^2}{\sqrt{1 - q^2/c^2}}$$

Enter the Atomic Nucleus

After any great scientific discovery, the question inevitably arises as to why the phenomenon or principle had not been discovered before. The answer is usually complicated, and several factors enter into play. One is that scientists often *had* bumped into it before, but had either ignored it, or misunderstood it, or incompletely described it. Such was indeed the case with the conversion of mass and energy. Another factor is that the existing scientific knowledge may be structured in a way to discourage people from seeing the phenomenon or principle as possible. That, too, was an element here, for mass and energy were viewed as entirely separate categories of nature, obeying separate laws. Finally, situations where the phenomenon or principle may show up in ways that scientists can easily explore may be rare and the effect tiny. That was also true, for conversions of mass into energy, or vice versa, are rarely observed in ordinary life. As Einstein wrote, "It is as though a man who is fabulously rich should never spend or give away a cent; no one could tell how rich he was."[25]

Was the wealthy man spending any money? And if so, where? In his "Energy Content" paper, Einstein expressed himself about the prospect far more cautiously than in his enthusiastic letter to Habicht. "Perhaps," Einstein wrote, "it will prove possible to test this theory" using substances such as radium that emit energy in the form of radiation. But the size of the effect, he remarked in a paper written shortly thereafter, would be "immeasurably small," and cited a calculation by Max Planck that radium's mass loss would be "outside the experimentally accessible range for the time being."[26] Still, Einstein added, "it is possible that radioactive processes will be detected in which a significantly higher percentage of the mass of the original atom will be converted into the energy of a variety of radiations than in the case of radium."

The discovery of the nucleus in 1911 did little for the moment to open any doors to testing the mass-energy concept, and matters remained unchanged for over two decades. But in 1932, two key developments made the mass-energy concept not only useful but indispensable for explaining aspects of the universe from its smallest to its largest dimensions—from atomic structure to stellar explosions. The first was the discovery of the neutron by British physicist James Chadwick. Physicists now had a good picture of the basic structural elements of the nucleus: protons and neutrons. Where did the energy come from that held them together? The clue was provided by a second key discovery of 1932, when British physicists John Cockroft and Ernest Walton used a new instrument in physics, a particle accelerator, to bombard lithium nuclei with protons, producing a nuclear transformation: the lithium nucleus plus the proton turned into two helium nuclei. Cockroft and Walton were able to measure the masses and energies of the initial states (lithium nuclei and protons) and of the final states (helium nuclei). They discovered a net mass loss, and energy gain—and established that, within experimental error, the mass loss was accounted for by Einstein's formula. The total inertial mass afterward, that is, was less by an amount equal to the increase in kinetic energy in the reaction divided by the speed of light squared. This was the first confirmation of Einstein's mass-energy equation, and it quickly became indispensable in atomic physics. The difference between the mass of particles inside and outside the nucleus was known as the "packing fraction," and the total mass difference between all such particles inside and outside is called the binding energy. Meanwhile, physicists were also learning that the energy of starlight came from mass-energy changing transformations in stellar interiors. In the 1930s, the concepts of packing fraction and binding energy made Einstein's equation a well-used tool of science, from atomic physics to astrophysics.

Physicists knew that even a small fractional conversion of mass to energy generated a lot more energy than any other kind of process they knew about. Still, the energy generated by any single nucleus—even if all of its binding energy were released—was far too minuscule for any practical purpose. For this reason, for the rest of the decade, nuclear energy seemed a distant, even ridiculous, prospect, the stuff of dreamers and fanatics. Almost to the end of the 1930s, nearly all physicists thought that the prospect of being able to release and control nuclear energy was far-fetched, even crazy. In 1921, Einstein was cornered by a young man proposing to produce a weapon based on $E = mc^2$. "Its foolishness is evident at first glance," Einstein replied.[27] In a 1933 interview, physicist Ernest Rutherford called the idea "moonshine." Einstein compared it to shooting in the dark at scarce birds. And in 1936, Danish physicist Niels Bohr, discussing instances when collisions between particles and nuclei that are so energetic that the nuclei explode, remarked that this would not "bring us any nearer to the much discussed problem of releasing nuclear energy for practical purposes." Bohr added, "Indeed, the more our knowledge of nuclear reactions advances the remoter this goal seems to be."[28]

By then, however, a series of events had already begun to unfold that would transform the world's appreciation for mass-energy conversions. The scientific and political events of this now-familiar story, with an international cast of characters, moved forward with a swiftness and drama that, even in outline, is still breathtaking well over half a century later.

Immediately after Chadwick's discovery of the neutron in 1932, physicists realized that the particle was an excellent tool for studying atomic nuclei. In the mid-1930s, as fascism grew in Europe, Italian physicist Enrico Fermi began bombarding elements of the periodic table with neutrons, going systematically from beginning to end, producing heavier, radioactive versions of each. When he reached the heaviest known element, uranium, he got strange results, and he thought he was creating brand-new, "transuranic" elements.

German scientists Otto Hahn and Fritz Strassman discovered that Fermi was wrong; adding neutrons to uranium actually produced lighter, already familiar elements. In December 1938 they mailed the news to Lise Meitner, a former co-worker who had fled Nazi Germany for Sweden. With her physicist nephew, Otto Frisch, Meitner realized that the bombardment was in effect splitting the nuclei, which Frisch named "fission" after consulting with a biologist. Frisch and Meitner sent a landmark paper on nuclear fission to *Nature*, which published it in February 1939—but by then Frisch had told Niels Bohr, who was about to embark on a boat for the U.S. Bohr and his companion broke the news to U.S. physicists the day they arrived, in mid-January, at the Princeton physics department journal club. The following week Columbia University physicists conducted the first fission experiment on U.S. soil, while word spread around the country like wildfire. Scientists first read about it not in journals but in newspapers. Most realized that fission—a process in which one uranium nucleus, in splitting and releasing neutrons able to split more nuclei—raised the possibility of a chain reaction, with massive numbers of nuclei splitting and releasing energy all at once—and thus of the possibility of a new, particularly terrifying type of bomb. This, just as Europe was on the brink of war.

In March 1939, Fermi (who meanwhile had fled Fascist Italy for the U.S., first to Columbia University in New York City and then to the University of Chicago) and other physicists began formally speaking to U.S. government officials about possible military applications. In July, two scientists visited Einstein at his summer home in Peconic, Long Island, to seek his help. "I never thought of that!" Einstein exclaimed after learning of the possibility of a chain reaction. Two weeks later, he signed an urgent letter to President Roosevelt informing him of "some recent work" that "leads me to expect that the element uranium may be turned into a new and important source of energy in the immediate future." In the last 4 months, the letter continued, the possibility has arisen of using uranium to set up a chain reaction. "This new phenomenon would also lead to the construction of bombs, and it is conceivable—though much less certain—that extremely powerful bombs of a new type may thus be constructed."

In September 1939, Nazi Germany invaded Poland. In October, Einstein's letter was presented to Roosevelt. In February 1940, the federal government awarded a $6,000 grant to study the phenomenon, dubbed the Manhattan Project. Several nations that were participants in the growing hostilities, including Germany, the Soviet Union, Japan, and Great Britain, began atomic bomb research. But events moved forward swiftly only in the U.S.

On December 2, 1942, less than a year after the project began serious work, the world's first controlled chain reaction took place at the Metallurgical Laboratory in a squash court in the west stands of the University of Chicago's football field, clinching the project's possibility (the news was communicated by the improvised code that "the Italian navigator [that is, Fermi] has just landed in the new world"). President Roosevelt then approved $400,000 for the project, leading to the construction of a huge isotope separation plant at Oak Ridge, Tennessee, and a plutonium production plant in Hanford, Washington. J. Robert Oppenheimer, the scientific head of the project, found a safe, remote site for the actual construction of the bomb atop a mesa in Los Alamos, New Mexico, and scientists began moving there in March 1943.

The Manhattan Project culminated in a test explosion at Alamogordo, New Mexico, on July 16, 1945. Scientists are used to witnessing new phenomena only in clinical lab conditions, but the Trinity test was different. That cold morning in the desert, Los Alamos scientists crouched down clutching pieces of welder's glass to protect their eyes. Suddenly a fireball erupted that was brighter than the sun, giving off heat that warmed their faces from 20 miles away. Slowly a white cloud rose tens of thousands of feet high, making some fear that they had unleashed force beyond their control, and reminding Oppenheimer (he said later) of scriptural passages about the apocalypse. It was, Abraham Pais wrote, "one of the most spectacular events in the history of the world."[29]

Three weeks later, on August 6, 1945, the first atomic bomb incinerated the Japanese city of Hiroshima. The next day, headlines all over the world revealed the existence of a new and particularly horrible type of bomb that, as *The New York Times* put it, "was the first time that Prof. Albert Einstein's theory of relativity has been put to practical use outside the laboratory."[30] On August 9, 3 days after the bombing of Hiroshima, another atomic bomb destroyed the city of Nagasaki.

The equation $E = mc^2$ had not played any direct role in the events leading up to the Manhattan Project, except as a key ingredient of the nuclear physics theory by which fission was understood. The atomic bomb, involving as it did the conversion of matter into energy via fission, was only an example of the equation $E = mc^2$—and a rare one in the comings and goings of life on earth—not an outgrowth of it. But the equation almost immediately became associated with it, partly thanks to *Atomic Energy for Military Purposes*, a report written by Princeton physicist Henry D. Smyth, an official of the Manhattan Project.

The Smyth report, as it was called, was released to the public on August 11, 2 days after the bombing of Nagasaki. "A weapon has been developed that is potentially destructive beyond the wildest nightmares of the imagination," Smyth wrote. It was created "not by the devilish inspiration of some warped genius but by the arduous labor of thousands of normal men and women working for the safety of their country." The report's intended audience was "engineers and scientific men" who might be able to "explain the potentialities of atomic bombs to their fellow citizens." But the book was a crossover hit, becoming a huge popular success and selling over a hundred thousand copies in its first 5 months.[31]

Right at the beginning, the Smyth report used $E = mc^2$ as a cornerstone for explaining the weapon. An early implication of relativity theory, it said, was the equivalence between mass and energy.

> To most practical physicists and engineers this appeared a mathematical fiction of no practical importance. Even Einstein would hardly have foreseen the present applications, but as early as 1905 he did clearly state that mass and energy were equivalent and suggested that proof of this equivalence might be found by the study of radioactive substances. He concluded that the amount of energy, E, equivalent to a mass, m, was given by the equation
>
> $$E = mc^2$$
>
> where c is the velocity of light.

The Smyth report was the mediating document through which nonscientists learned about the Manhattan Project. It more than any other single document made $E = mc^2$ an emblem of atomic energy and weaponry.

Celebrity Status

Ethnographers say that when two cultures interact, they do not meet all of a piece but through "congeners" through which certain members of one culture look at, try to understand, and respond to the other. Congeners can include artifacts, rituals, practices, and art; fear, fascination, and exoticism usually play a role. A congener is like a little lens that allows the members of the one culture to approach the other culture in a focused way, to get an introductory grip. A congener is thus more than a symbol or logo of the other culture, but guides and disciplines curiosity and fascination into a first interaction with it.

The equation $E = mc^2$ served as a congener in this sense, between a public anxious for information about atomic energy and the scientific developments that made it possible. In the process, it grew into even more—a symbol of physics, of science, even of knowledge—to the point where it acquired a mythic status.

The French intellectual Roland Barthes once wrote an essay on Einstein in which he noted that, while photographs of Einstein often show him next to a blackboard covered with impenetrable symbols and equations, cartoons of him often portray him, chalk in hand, next to a clean blackboard on which he has just written down this particular formula as if it just came to him out of the blue. Barthes observed of the symbolic character of this equation that it restores the image of "knowledge reduced to a formula . . . science entirely contained in a few letters." It has become a Gnostic image: "the unity of nature, the ideal possibility of a fundamental reduction of the world, the unfastening power of the word, the age-old struggle between a secret and an utterance, the idea that total knowledge can only be discovered all at once, like a lock which suddenly opens after a thousand unsuccessful attempts." Barthes's essay helps explain the transformation of this equation from a scientific tool into a congener.

Einstein himself began to use the equation in its now-famous, simplified form. In April 1946, the first issue of a new popular magazine, *Science Illustrated*, appeared with an article written by Einstein entitled "$E = mc^2$." "It is customary," he wrote, "to express the equivalence of mass and energy (though somewhat inexactly) by the formula $E = mc^2$."[32]

Then, on July 1, 1946, less than a year after the explosions over Hiroshima and Nagasaki, and just shy of 41 years after it first appeared in its initial form, $E = mc^2$ made the cover of *Time* magazine. The issue coincided with an atomic test in the South Pacific. The *Time* cover juxtaposed a portrait of the now white-haired, 66-year-old Einstein, whom it called a "shy, almost saintly, childlike little man," next to a fiery mushroom cloud rising above the hulks of warships. Red flames at the base gave way to orange and purple on the column, topped by a gray mushroom cap. On it was inscribed a now-infamous equation: $E = mc^2$. Now it was a celebrity.

INTERLUDE

Crazy Ideas

In order to understand this we need some crazy idea. Has anyone a crazy idea?

—Niels Bohr

The scientist and science writer Jeremy Bernstein writes that he occasionally fantasizes the following:

> It is the year 1905 and I am a professor of physics at the University of Bern. The phone rings and a person I have never heard of identifies himself as a patent examiner in the Swiss National Patent Office. He says that he has heard I give lectures on electromagnetic theory and that he has developed some ideas which might interest me. "What sort of ideas?" I ask a bit superciliously. He begins discussing some crazy sounding notions about space and time. Rulers contract when they are set in motion; a clock on the equator goes at a slower rate than the identical clock when it is placed at the North Pole; the mass of an electron increases with its velocity; whether or not two events are simultaneous depends on the frame of reference of the observer; and so on. How would I have reacted?[1]

Bernstein's thought experiment highlights an occupational side-effect of writing about science: letters from strangers bearing crazy ideas. Once upon a time these letters arrived in brown envelopes in crabbed handwritten script; today they are emailed with links to flashy Web pages. The usual subjects are astrophysics, cosmology, unification theories, and the overthrow of Western science. Einstein is often invoked, either as emblematic of mainstream science (as the author's archenemy) or of the misunderstood loner outside it (as the author's precursor). One or more of the following are generally mentioned: gravity, electromagnetism, and planetary orbits; the most primitive and readily recognizable versions mention also psychic phenomena, astrological signs, herbal remedies, the stock market, baseball scores, and rock lyrics. Many crackpot letters explode into *italics*, **boldface**, and CAPITAL LETTERS in a way reminiscent of far-right newsletters and software licensing agreements. Some authors warn of a government or scientific conspiracy to suppress their ideas; others are generously allowing you into the Fold.[2]

Few who receive crazy-idea letters reply. It's assumed to be counterproductive and perhaps even dangerous, reinforcing the authors' sense of being misunderstood and inviting more urgent appeals. The recipients quickly peruse the letters, then toss them into a "crackpot letter" drawer. Hardly anyone I know throws them out.

Why not?

One of my colleagues compares his "crackpot letter" drawer to the neighborhood art show; if you're thorough and patient enough you might find *something* of value, but the search would take so long you never do it. Others offer psychological explanations: We admire and even envy the authors of crackpot letters for their energy and zeal. We feel a closet affinity—don't we all feel in possession of misunderstood truth? Still darker, we're thrilled to read them—it's like watching a mental train wreck. Yet other colleagues keep them because, they tell me, "You never know . . ."

Crazy-idea letters have interesting philosophical, psychological, and social dimensions. It's much harder than it seems to characterize crazy-idea letters and their appeal. Great scientists, too, have gone weird: recall the obsessions of Einstein with unified field theories and Pauling with vitamin C. And aren't scientists fond of and even dependent on "crazy ideas"? Haven't we all heard the famous story about how Wolfgang Pauli charged Niels Bohr with holding a crazy theory, and how Bohr replied that the trouble was his theory was "not crazy enough" and called for a "crazier idea"?

Finally, several cautionary tales suggest we should not be too confident of our ability to recognize crazy-idea letters.

Cautionary tale 1 is the story of 25-year old Srinivasa Ramanujan, an Indian who in 1913 sent correspondence to several British mathematicians. Like several others, G. H. Hardy tossed it aside initially as the work of a crackpot, then read it, realized its genius, and soon invited Ramanujan to England, where he became recognized as one of the leading mathematicians of all time. Our own prejudices, rather than content, can determine whom we deem a nut.

But cautionary tale 2 is the story of Nicholas Christofilos, a Greek electrical engineer at an elevator installation company whose hobby was particle accelerators. In 1949, he sent a manuscript proposing a novel scheme to physicists at Berkeley, who wrote back pointing out flaws. Christofilos incorporated corrections, applied for U.S. patents, and sent the revised scheme back to the Berkeley physicists, who this time simply ignored it. In 1952, reading that U.S. physicists had "discovered" a new accelerating method identical in principle to his own, he contacted a legal firm and got his priority recognized. I once asked a physicist who had seen Christofilos's original papers why they had been ignored. "The first violated Maxwell's equations," he said, shrugging without elaborating, his body language indicating that this was equivalent to mentioning psychic phenomena, and that he therefore needed to make no apology for ignoring the rest. Bad physics does not necessarily make a crackpot.

Jeremy Bernstein insists that, had he seen a copy of his correspondent's recently published paper, "The Electrodynamics of Moving Bodies," he would have been able to tell that this was not a crank paper, and cites two clues. The first clue is "connectivity," or the fact that the theory gave the same answers as Newtonian mechanics when the speeds of bodies are low compared with that of light. Crank theories "usually start and end in midair" without genuinely connecting with the existing body of scientific knowledge. The second clue is the presence of testable predictions.

I would add two more clues. The first involves the way the authors handle equations. Crazy-idea letters almost always include either no equations or a small number treated like fetish objects. Equations in such letters, indeed, generally illustrate Barthes' observation about the Gnostic fantasy of knowledge reduced to a formula. The equations appear unaccompanied, as if they were independent nuggets of truth. They are treated as aphorisms, encapsulizing entire philosophies in miniature. The role of the equation in the paper is like that of a musical instrument that someone carries around without ever playing. In genuine scientific papers, by contrast, equations hardly ever occur unaccompanied, but are embedded in a sequence as coparticipants in an extensive logical argument, fragments of an extensive intellectual edifice to which they owe their very existence, only a small part of which is reproduced on the page. The equations, that is, are not treated as standing completely on their own.

But I think that the most important clue to a crackpot is the lack of an engaged attitude coupled with playfulness—the sort of attitude that Einstein displayed in his fear he mentioned to Habicht that God was leading him around by the nose, but in his willingness to go along.

As evidence I submit the following story. It took place in September 1946 in New York City at one of the first postwar annual meetings of the American Physical Society. At one session, the presentation by the young Dutch theorist Abraham Pais, who was struggling to explain the strange behavior of a puzzling, recently discovered new particle, was interrupted by Felix Ehrenhaft, an elderly Viennese physicist. Ever since 1910, Ehrenhaft had been claiming to have evidence for the existence of "subelectrons," charges whose values were smaller than the electron's, and his efforts to advance his claims had long ago exhausted the patience of the physics community. Now approaching seventy, Ehrenhaft was still seeking an audience, and approached the podium demanding to be heard.

A young physicist named Herbert Goldstein—who told me the story—was sitting next to his mentor and former colleague from the MIT Radiation Laboratory, Arnold Siegert. "Pais's theory is far crazier than Ehrenhaft's," Goldstein asked Siegert. "Why do we call Pais a physicist and Ehrenhaft a nut?"

Siegert thought a moment. "Because," he said firmly, "Ehrenhaft *believes* his theory."

The strength of Ehrenhaft's conviction, Siegert meant, had interfered with the normally playful attitude that scientists require, an ability to risk and respond in carrying forward their dissatisfactions. (Conviction, Nietzsche said, is a greater enemy of truth than lies.) What makes a crackpot is not simply our prejudices, nor necessarily the claim, but our recognition of the disruptive effects of the author's conviction. For conviction tends to wipe out not only the dissatisfaction but also the playfulness, the combination of which produces such a powerful driving force in science.

CHAPTER EIGHT

The Golden Egg: Einstein's Equation for General Relativity

$$G_{im} = -\kappa(T_{im} - \tfrac{1}{2}g_{im}T)$$

DESCRIPTION: Space-time tells matter how to move, matter tells space-time how to curve.
DISCOVERER: Albert Einstein
DATE: 1915

[O]ne of the greatest achievements of human thought . . .

—J. J. Thomson

The theory appeared to me then, and it still does, the greatest feat of human thinking about nature, the most amazing combination of philosophical penetration, physical intuition, and mathematical skill. But its connections with experience were slender. It appealed to me like a great work of art to be enjoyed and admired from a distance.

—Max Born

Einstein's field equation of general relativity, expressing the curvature of space-time, is not as instantly recognizable as his formula of special relativity, expressing the interchangeability of mass and energy. But it, too, vaulted to public awareness under dramatic circumstances.

The date was November 6, 1919, the place a chamber of the Royal Society building in London. The room resembled a small church interior, with rows of pews on either side of a central isle and a row of columns lining the walls. In the back was an anteroom for an overflow crowd. On one wall hung a portrait of the Royal Society's most famous member, Sir Isaac Newton.

The event was a joint meeting of the members of the Royal Society of London and the Royal Astronomical Society. The audience had come to hear reports of data collected during an eclipse that had taken place some 6 months earlier, on May 29. Scientists had taken photographs of the stars during the eclipse, trying to see whether the starlight had been bent as it passed around the sun. Some of those in attendance had alerted the press to the importance of the occasion. The *Times* of London, in an extensive story, declared, "The greatest possible interest had been aroused in scientific circles by the hope that rival theories of a fundamental physical problem would be put to the test, and there was a very large attendance of astronomers and physicists."[1] One of the rival theories was Albert Einstein's, whose idea that space was "curved" was part of his equation of general relativity; this idea in turn implied the bending of starlight around the sun. British physicist J. J. Thomson, a grand old man of British physics who had discovered the electron, presided. "This is the most important result obtained in connection with the theory of gravitation since Newton's day," Thomson announced, and described the result as "one of the highest achievements of human thought."[2] Philosopher Alfred North Whitehead, an audience member, later wrote:

> The whole atmosphere of tense interest was exactly that of the Greek drama: we were the chorus commenting on the decree of destiny as disclosed in the development of a supreme incident. There was dramatic quality in the very staging:—the traditional ceremonial, and in the background the picture of Newton to remind us that the greatest of scientific generaliza-

> tions was now, after more than two centuries, to receive its first modification. Nor was the personal interest wanting: a great adventure in thought had at length come safe to shore.[3]

Special relativity, too, had been an adventure, but one in which many members of the scientific community—including FitzGerald, Lorentz, Poincaré, and many others puzzled by the contradiction between Newton and Maxwell—had participated. General relativity was different. Einstein embarked on it virtually alone. For 7 years his pursuit of the topic had followed a mazelike path, in which his course was helped by doors that unexpectedly opened, or was blocked by dead ends that forced him to retrace steps and undo years of work. Only at the very end did others realize that a journey of drama and extraordinary significance had taken place.

Solo Journey

The path to general relativity, too, was initiated by a thought experiment, which occurred to Einstein sometime in November 1907. He once wrote:

> I was sitting in a chair in the patent office at Bern when all of a sudden a thought occurred to me: If a person falls freely he will not feel his own weight. I was startled. This simple thought made a deep impression on me. It impelled me toward a theory of gravitation.[4]

He referred to this insight as "the happiest thought of my life."[5]

This thought experiment was an ambitious extension of the thought experiments associated with special relativity. Those had involved situations of uniform motion, and the point was the inability to tell whether they were moving: everything in that smooth and cushy railroad car, for instance, acts exactly the same when it is in motion as when it is at rest. The new thought experiment involved accelerated motion: imagine that the railroad car is lifted high, is released, and falls to the ground. Einstein realized that you could not tell whether you were falling in a gravitational field, or out in empty space in the absence of any gravitational field. If you let go of an object—keys, a ball, coins—they would stay put as if at rest. Einstein called it his "happiest thought" because he realized that that lack of an ability to distinguish the two situations was important.

Today, thanks to countless news clips of weightless conditions in spacecraft, in high-altitude planes, and so forth, we find this thought much less startling than did the 28-year-old Einstein. The idea that free fall in a gravitational field was indistinguishable from the absence of force suggested to him that the presence of gravitation was identical to the presence of acceleration—that is, force—itself. For you could reverse the situation: If the railroad car were on the ground, could you tell whether you were in a gravitational field or being accelerated? These thoughts focused Einstein's dissatisfaction, and he set out to explore the implications of what not being able to tell the difference might mean.

This thought experiment involved a different sort of dissatisfaction than the one that had led to special relativity. Instead of bringing to a head a contradiction born from trying to combine two complete systems—Newton's and Maxwell's—this one was produced by an apparent identity between two things thought to be vastly different. It is as if, out of tradition and habit, we were used to dealing with two different governmental offices for different things, discovered that their actions were identical, and then had to figure out how this could possibly be true. The two things that seemed different were inertial mass and gravitational mass. According to Newton, gravitation was a special kind of force that tugged at heavier objects more strongly than at lighter objects—but by sheer coincidence, the inertial mass of the heavier ones made them resist the tug by just the right amount so that everything accelerated at the same rate. Einstein thought: let's assume that this is no coincidence, and see what happens.

Another way of putting this is to say that he was probing for an even deeper covariance than the one he had discovered in his earlier theory of relativity. Covariance is simply part of what we mean by objectivity: to say that something is a real part of the world is to say that it looks different from different "angles"—including not only things like lighting conditions but also spatial locations and speeds—in a way that you can spell out precisely through "transformations." Covariance thus seems to draw a difference between how a thing appears to us, and what it really is. In Einstein's 1905 theory, he had discovered transformations that would make the same description apply no matter how the object moved—so long as it was moving uniformly—even though the object would "look" different if it were traveling at close to the speed of light. If something did not behave that way—did *not* look different if it were traveling at close to the speed of light—we were entitled to say it was not a real object, not a part of our world. Now Einstein was trying to extend covariance to accelerated systems: How would the description of a "real" object change if the reference frame were accelerated? This would lead him to such a radical reformulation of his work that it would be called "general relativity" to distinguish it from its precursor, "special relativity."

First Step: The Principle of Equivalence (1907)

Einstein's thought experiment inaugurating the solo journey had occurred to him while working on the paper summarizing his theory of relativity for the *Jahrbuch der Radioaktivität*. He added a final, ten-page section to the paper with a "novel consideration." Up to now, he wrote, he has applied the principle of relativity to uniformly moving systems, but: "Is it conceivable that the principle of relativity also applies to systems that are accelerated relative to each other?" What if there were two systems, one accelerated at a certain rate and the other at rest in a uniform gravitational field exerting the same force? As far as we know, he says, the physical laws of the two systems are the same, and "we shall therefore assume the complete physical equivalence of a gravitational field and a corresponding acceleration of the reference system."[6] Einstein wrote that he did not know whether this "equivalence principle" is true—he just means to see what would happen if it were.

The next few pages contain many key ingredients of what would become general relativity. He draws surprising conclusions, such as that people higher in a gravitational field will perceive clocks farther down (toward the source) as moving slower, and that gravitational fields affect the path of light—though "unfortunately, the effect of the terrestrial gravitational field is so small . . . that there is no prospect of a comparison of the results of the theory with experience."[7] Still, he hoped for testable predictions.

Einstein, for instance, had his eye on the longstanding problem of Mercury's orbit. In the mid-nineteenth century, astronomers had noted that the point at which it made its closest pass by the sun—called the perihelion—was not staying put but slowly moving around the orbit. At first astronomers assumed that this was due to the gravitational influence of other planets, but when these influences were all scrupulously added in, a tiny amount—43 seconds of arc per century—still could not be accounted for. The discrepancy was not small enough to ignore, but not big enough to cast doubt on the Newtonian system within which it appeared as a problem. Fixes were proposed. Some involved postulating phenomena such as a hidden planet named Vulcan or a lenslike layer of nebulous matter near the sun, but these could not be found. Another approach was to tinker with Newton by introducing slight modifications of the inverse square law, but these had undesirable side effects. For half a century the precession of Mercury's orbit had been one of the great unsolved mysteries of astronomy.

Einstein realized that this final piece of the precession was something his theory might explain from first principles thanks to certain terms in his theory that were absent from Newton's, so that astronomers would not have to hang their hopes for a predictable universe on undiscovered phenomena, tinkering with constants, or adjusting formulas. "At the moment," he wrote his friend Habicht on Christmas Eve 1907, "I am working on a relativistic analysis of the law of gravitation by means of which I hope to explain the still unexplained secular changes in the perihelion of Mercury."[8] But for several years these changes remained mysterious.

Einstein wrote nothing about gravitation from 1907 to 1911. For one thing, there were personal disruptions; the birth of another son in 1910, a move to Prague with his wife Mileva and two children in March of 1911. His new jobs—associate professorships in Zürich at the Eidgenössiche Technische Hochschule (ETH), then in Prague in 1911—brought more pressures. And he became absorbed in the problems of quantum theory, in the course of which he showed the world of physics how to banish what Kelvin had called "Cloud No. 2": the difficulty of applying the Maxwell-Boltzmann theory to certain experimental results, which Einstein showed to be due to effects relating to the quantum idea.

By the time Einstein returned to gravitation, a promising new door had opened in the maze, though he passed it by for the moment. It had been opened by a former ETH professor of his, Hermann Minkowski (1864–1909). Minkowski had not been especially enamored of his student, whom he had once called a "lazy dog"; and on seeing Einstein's 1905 paper remarked, "Imagine that! I would have never expected such a smart thing from that fellow."[9] But Einstein turned out to be Minkowski's dream student, one who borrowed, assimilated, and transformed what he had learned, instructing his teacher.

The door Minkowski opened, inspired by Einstein's theory of special relativity, was a mathematical approach to space and time that put them on equal footing. Minkowski described objects as having not only an *x*, *y*, and *z* of position, but also a *t* corresponding to their position on a fourth axis, time. This new approach, that is, conceived of objects as moving along a time "line" in just the way they moved along spatial lines. An object that stayed in the same place would be represented by a straight line, with only its position on the t-axis changing. An object moving uniformly along the x-axis would be represented by a curved line, for it had not only moved along the x-axis but also uniformly along the t-axis. More complexly moving objects followed more complex routes. The distance between an object in one position and another had four terms rather than three; it was like adding yet another "side" to the Pythagorean theorem corresponding to the time difference between the two events. While the Pythagorean theorem for three-dimensional Euclidean space is $s^2 = x^2 + y^2 + z^2$, where s^2 is the length that is invariant for all observers, the extension of the theorem to "space-time" now meant that it was $s^2 = x^2 + y^2 + z^2 - (ct)^2$.

Minkowski's formulas extended covariance further, by including in the conditions for objective existence an object's behavior in time. A description of something that gives only the x, y, and z of position without a position in time is like a snapshot of a person compared to a video—only a partial view. In space-time, a full description of an object is more like a video, for it charts the object's spatial location *and* time at once. To generate the formulas, Minkowski made use of mathematical tools, now called "tensors," that translate sets of quantities from one coordinate system to another. Tensors are mathematical objects that are "ranked" based on their indices, which is a measure of how complex they are. A tensor of rank 0 is a constant, of rank 1 is a vector, and of rank 2 is a set of complex matrices with multiple components that can be used to translate one set of coordinates into another. Rank 2 tensors allowed Minkowski to bring space and time together, in the process redefining them as thoroughly as Einstein had simultaneity. In 1908, Minkowski gave a talk in Cologne that began:

> Gentlemen, the views of space and time which I wish to lay before you have sprung from the soil of experimental physics, and therein lies their strength. They are radical. Henceforth space by itself, and time by itself, are doomed to fade away into

mere shadows, and only a kind of union of the two will preserve an independent reality.[10]

But for the moment Einstein passed by this door in the maze, dismissing his teacher's work as "superfluous learnedness."[11] He would be forced to return.

In 1911, now in Prague with Mileva and his children, Einstein published what amounted to a sequel to his 1907 paper, entitled "On the Influence of Gravitation on the Propagation of Light."[12] In an earlier paper, he began, I addressed the question of whether gravitation affects light. But "my former treatment of the subject does not satisfy me . . . because I have now come to realize that one of the most important consequences of that analysis is accessible to experimental test." Two tests, in fact. According to the equivalence principle, "one can no more speak of the *absolute acceleration* of the reference system than one can speak of a system's *absolute velocity* in the ordinary theory of relativity." This had implications for how light behaves in a gravitational field. One, which he discussed in section 3, was the existence of a *gravitational red shift*, that light emitted by sources "further down" in such a field—at the sun's surface, for instance—would be shifted slightly toward the red when compared to similar sources "higher up," such as on the earth. Measuring the difference, though difficult, would be one important test of general relativity.

A second test, discussed in section 4, was the *bending of starlight*. For "rays of light passing near the sun experience a deflection by its gravitational field, so that a fixed star appearing near the sun displays an apparent increase of its angular distance from the latter, which amounts to almost one second of arc." Suppose astronomers took photographs of the stars in the background during a total solar eclipse, and compared those with photographs of the same stars without the sun. If, as Einstein suspected, the starlight bent as it brushed past the sun, the two photographs would be different, and the stars would appear farther from the sun in the first picture. He now predicted how much.

> A ray of light going past the sun would accordingly undergo deflection to the amount of $4 \cdot 10^{-6} = .83$ seconds of arc. . . . As the fixed stars in the parts of the sky near the sun are visible during total eclipses of the sun, this consequence of the theory may be compared with experience. . . . It would be a most desirable thing if astronomers would take up the question here raised. For apart from any theory there is the question whether it is possible with the equipment at present available to detect an influence of gravitational fields on the propagation of light.

This value, shortly corrected to .87 seconds of arc, simply reflects the fact that light, because it has mass according to $E = mc^2$, is as affected by gravitation as stones and apples, and is based on Newtonian principles; in fact, it is known as the "Newtonian value."

In this 1911 paper, Einstein still could say nothing about the still-mysterious *precession of Mercury's perihelion*, which would become the third key prediction of general relativity. But the first two predictions—especially the bending of starlight—set in motion events that would culminate 8 years later in the famous meeting, and make headlines all over the world. For Einstein began to agitate for astronomers to look into these predictions.

In August 1911, he sent his "Influence of Starlight" paper to an astronomer at the University of Utrecht who had authored a paper about the solar redshift; Einstein wrote that he had arrived at the "somewhat daring" conclusion that "the gravitational potential difference" might be the cause. "A bending of light rays by gravitational fields also follows from these arguments." If something else is responsible, Einstein added, "then my darling theory must go in the wastebasket."[13] That month, too, Einstein sent his paper to Erwin Freundlich, a young Berlin astronomer who would become relativity's advocate among astronomers. Freundlich offered to look for the influence of Jupiter on starlight, and to investigate photographs taken during eclipses. Jupiter proved not massive enough to bend starlight to any detectable degree. "If only we had an orderly planet larger than Jupiter!" Einstein lamented. "But Nature did not deem it her business to make the discovery of her laws easy for us."[14] Freundlich also set out to see if stellar deflections could be detected from photographs taken of past eclipses, but this proved impossible. And he initiated an expedition to photograph an eclipse in Brazil in October 1912. But the expedition was rained out, one organizer wryly commenting that "we . . . suffered a total eclipse instead of observing one."[15]

Undaunted, Freundlich set out himself on an expedition to Russia—one of several—for an eclipse that would take place in August 21, 1914. Einstein excitedly wrote Ernst Mach in June 1913, "Next year, during the solar eclipse, we shall learn whether light rays are deflected by the sun, or in other words, whether the underlying fundamental assumption of the equivalence of the acceleration of the reference system, on the one hand, and the gravitational field, on the other hand, is really correct."[16] That fall, he wrote astronomer George Hale asking if a powerful enough telescope could detect stars near the sun during the day, so that he would not have to wait for a solar eclipse; but the eminent astronomer dashed his hopes, and Einstein resigned himself to waiting for the August 1914 eclipse. But that expedition, too, met disaster, for political as well as meteorological reasons. The assassination of Archduke Francis Ferdinand on June 28 was followed by Austria-Hungry's invasion of Serbia, and Germany's declaration of war on Russia on August 1. Freundlich's expedition was an early war casualty. He was arrested—though soon ransomed and able to return to Berlin—and his equipment impounded. Other expeditions were stymied by bad weather.

Einstein tried to help where he could, and endured the setbacks with some impatience. But the collapse of the attempts proved beneficial to him in the long run. For by the time of the Russian expedition Einstein had entered a different part of the maze, causing him to revise his predictions.

Second Step: Geometry of Space-Time

In the meantime, thanks in part to his struggles with the bending of light in a gravitational field, Einstein had realized an even deeper implication of the principle of equivalence—that he would have to link his equations with different geometries of space. For a space to have a geometry does not mean that the space is literally curved—something can only be curved with respect to something else taken to be straight—but rather has to do with the way that the measurements of a path of something going through that space (a beam of light, say) add up. If light were bent by a gravitational field, the measurements of its trajectory add up in a way that corresponds to the mathematics of a certain geometry. This involved treating gravitation not as a force, that is, as something that exerts a tug, but as a property of space itself, as having a structure or architecture that must be followed by things passing through it. Writing equations without such architecture—without the geometry of space—Einstein wrote, would be like "describing thoughts without words."[17] Moreover, he also now realized the virtues of his old (and now deceased) teacher Minkowski's work, and how the use of tensors would vastly simplify the task he was undertaking. Accordingly, he retraced his steps in the maze and entered the door Minkowski had opened. But there he found himself overwhelmed by the task of finding tensors to work for curved, non-Euclidean, geometries.

Fortunately, he knew where to turn. In summer 1912 Einstein had moved from Prague to Zürich, recruited by his former companion and classmate Marcel Grossmann. In school, Einstein used to borrow Grossmann's mathematics lecture notes so he could concentrate on physics; now, Grossmann was dean of the ETH's mathematics-physics section. "Grossmann," Einstein wrote in desperation, "you must help me or else I'll go crazy!"[18] Grossmann then told him about non-Euclidean geometries that mathematicians had produced—by Bernhard Riemann (who had a twenty-component curvature tensor), Gregorio Ricci-Curbastro (who had developed a contracted version with ten components), and Tullio Levi-Cività—and explained to him about tensor calculus. With Grossmann's help, Einstein set out to develop a generalized, Minkowski-like tensor for a four-dimensional surface to represent the gravitational field. In August he excitedly wrote to one friend, "The work on gravitation is going splendidly. Unless I am completely wrong, I have now found the most general equations."[19] In October he wrote to another:

> I am now working exclusively on the gravitation problem and believe that I can overcome all difficulties with the help of a mathematician friend of mine here. But one thing is certain: never before in my life have I troubled myself over anything so much, and I have gained enormous respect for mathematics, whose more subtle parts I considered until now, in my ignorance, as pure luxury! Compared with this problem, the original theory of relativity is child's play.[20]

The next year, 1913, Einstein and Grossmann published an article, "Outline of a General Theory of Relativity and a Theory of Gravitation," that comes within a "hair's breadth" of the final equations.[21] It is in two parts, part I on physics by Einstein, and part II on mathematics by Grossmann. In writing it, they encountered hints that general covariance was impossible. This finding greatly disturbed Einstein, who called it an "ugly dark spot" in the theory.[22] He abandoned his hopes for general covariance of the field equation, and embarked on what would turn out to be a dead end from which it would take him 2 years to extricate himself.

Early in 1914, Einstein left Zürich for a new position in Berlin, leaving behind not only Grossmann but also Mileva, their marriage broken. At a farewell dinner thrown by an old ETH friend, Einstein expressed some apprehension, and complained of being treated as a "prize hen." As he told the person who walked him home, "I don't even know whether I'm going to lay another egg."[23]

Third Step: Covariant Equations

Once in Berlin, Einstein threw himself into work on general relativity, ignoring everything else; entering Einstein's study one day, Freundlich saw a meat hook hanging from the ceiling, from which the scientist impaled letters that he had no time to read.[24] Between October and November 1915, Einstein finally realized his mistake, and turned back to seeking a formulation that was generally covariant. That time, he wrote a friend, would be "one of the most stimulating, exhausting times of my life."[25] On November 4, Einstein told the Prussian Academy that he had "lost trust in the field equations" that he had earlier reported. He explained that he had returned to the "demand of general covariance" for the field equations, "a demand from which I parted, though with a heavy heart, three years ago," and had produced a truly covariant theory. "Nobody who really grasped it can escape from its charm."

Sometime during the next 2 weeks, he made a discovery that, one biographer wrote, was "by far the strongest emotional experience in Einstein's scientific life, perhaps in all his life."[26] He noticed that his new theory perfectly accounted for the precession of Mercury's perihelion. The prediction he had mentioned in 1907, and so diligently worked on, now popped right out of his new theory without any extra hypotheses or assumptions. "I was beside myself with joy and excitement for days," he wrote to Lorentz.[27] He told another friend he had heart palpitations, and yet another that something "snapped" inside him.[28]

On November 18, Einstein announced to the members of the Prussian Academy that his new, covariant gravitational field equations account for Mercury's orbit, and also—something he realized for the first time—that the theory of curved space-time also implies a prediction of the deflection of starlight around the sun of twice the amount he had predicted earlier. "This theory, however, produces an influence of the gravitational field on a light ray somewhat different from that given in my earlier work. . . . A light ray grazing the surface of the sun should experience a deflection of 1.7 sec of arc instead of 0.85 sec. of arc." The old Newtonian value was .85 seconds of arc; the new "Einsteinian value" was 1.7 seconds of arc.[29]

And, on November 25, in a talk called "The Field Equations of Gravitation," he wrote it down as follows, in the familiar way:[30]

$$G_{im} = -\kappa(T_{im} - \tfrac{1}{2}g_{im}T)$$

though it is sometimes written with *R*s instead of *T*s, and with additional terms.[31]

The equation comes in two parts. The left-hand side refers to a set of terms that characterize the geometry of space. The right-hand side refers to a set of terms that describe the distribution of energy and momentum. The left is the geometry side, the right is the matter side. As physicist John Wheeler liked to describe it, reading from left to right is space-time telling mass how to move; reading from right to left is mass telling space-time how to curve. Though the equation predicts only the slightest of deviations experimentally from the world as we know it—the positions of a few stars out of place—it amounts to a conceptual revolution from the world of Newton. In this new world, there is no absolute time nor space, and gravitation is not a force—not a tug between one object and another—but a property of space and time.

Einstein was utterly confident that it was right. He sent physicist Arnold Sommerfeld a postcard: "You will become convinced of the General Theory of Relativity as soon as you have studied it. Therefore I will not utter a word in its defense."[32] But while the core of general relativity was now clear to Einstein, it was still a maze to nearly all others. "[T]he basic formulas are good, but the derivations abominable," Einstein wrote to Lorentz.[33] Early in 1916, therefore, Einstein sat down to compose a logical route that others could follow. The result was a fifty-page paper published in a March issue of the *Annalen der Physik*, "The Foundation of the General Theory of Relativity." The paper was such a stunning success that it was reprinted in booklet form, ran through several printings, and was translated into English. The final section states the three experimental predictions made by the theory:[34]

> the spectral lines of light reaching us from the surface of large stars must appear displaced towards the red end of the spectrum.
>
> a ray of light going past the Sun undergoes a deflection of 1.7 [seconds of arc]; and a ray going past the planet Jupiter a deflection of about .02 [seconds of arc].
>
> The orbital ellipse of a planet undergoes a slow rotation. . . . Calculation gives for the planet Mercury a rotation of the orbit of 43 [seconds of arc] per century, corresponding exactly to astronomical observation (Leverrier).

The first prediction was too difficult at the time to confirm, and the third was already contained in the theory when it first appeared. But the second prediction—that part having to do with starlight passing the sun—looked like it could be tested.

With a little help from nature.

For experimental science is the art of getting things you understand to tell you things you do not understand. You roll balls down an incline and time them. You time the swing of a pendulum. You measure how oil droplets behave in an electric field. What's truly marvelous—the wonder of science—is that you get out of such events more than what you put into them. In some cases, you can stage these events as command performances, in the laboratory and entirely under your control. In other cases, you have to wait for nature to set the stage for you.

And an eclipse is one of these grand cosmic performances.

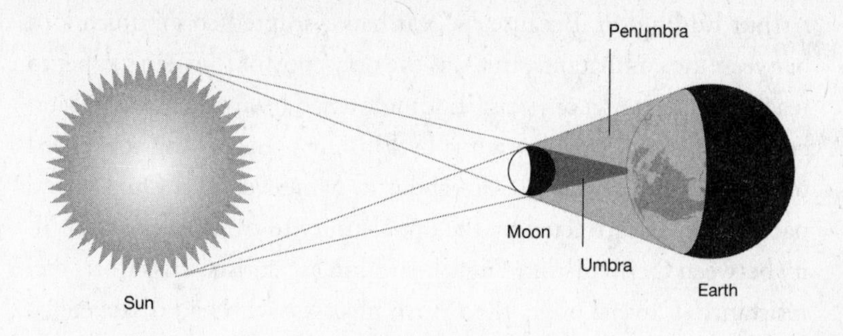

Cosmic Performance

An alien, watching with a powerful telescope from a great distance, would see the earth and the moon, bathed in light from the sun, cast huge cone-shaped shadows as they revolve about each other. These motions can be predicted exactly, thanks to Newton's laws. Every so often the earth or the moon enters the shadow cast by the other—sometimes completely, sometimes only partially. As seen from the earth, by an incredible coincidence, the moon's apparent size is the same as the sun's, so the moon completely blots out the sun's light, throwing everything in the shadow-cone into darkness, bringing back the stars whose light is ordinarily blotted out during the day. Such eclipses thus set the stage for being able to tell whether this starlight is bent as it passes by the sun.

Einstein was anxious and excited about the prospect of testing his theory. How could he not have been? It was more than a formula—it was a proposal that the fundamental structure of the universe was different, and stranger, than human beings had ever thought. He had been frustrated by the setbacks of the earlier expeditions. But these had spared him the embarrassment of testing a result he would have had to revise.

After his monumental 1916 "General Theory" paper outlining the trio of predictions, Einstein began actively promoting again the cause of testing the theory during an eclipse. He sent a copy of the paper to the Dutch astronomer and physicist Willem de Sitter, who in turn passed it to the secretary of the Royal Astronomical Society,

Arthur Eddington. Because the war had disrupted communications between the combatants, this was the only copy of Einstein's paper to reach England. As a physicist, Eddington had been vexed by the Mercury discrepancy, and intrigued by Einstein's theory, and proceeded to write several review articles about it. Moreover, as a Quaker and pacifist, he was attracted by the opportunity to overcome the hostility between German- and English-speaking scientists. His efforts met resistance. Oliver Lodge, the British physicist who had discovered a form of electromagnetic waves simultaneously with Hertz, and who continued to believe in the existence of ether, claimed that ether drift could account for Mercury's perihelion. An American opponent of Einstein named Thomas J. J. See, who insisted that gravitation was a real physical force, wrote that "the whole doctrine of relativity rests on a false basis and will someday be cited as an illustration of foundations laid in quicksand."[35] Nevertheless, Eddington persisted, and began to drum up interest in an eclipse expedition in this fascinating and fundamental theory.[36] Some recent social constructivist work, which tends to see all interest as self-interest, has drawn attention to Eddington's efforts to garner scientific, media, and public attention to relativity, with the implication that Eddington was engaged essentially in manipulation, public relations, and self-promotion.[37] But it is only natural that Eddington was excited by this work and wanted to share his excitement—and that the public and the media responded positively to this news of potentially fundamental significance. To think otherwise is to have an infantile view of scientists, and of the public. Had Eddington *not* wanted to share his excitement at scientific work that hinted at a revolutionary picture of the world—*that* would have made him pathological.

The latest eclipse expeditions fared no better than the earlier ones. The war scotched plans to mount an expedition to Venezuela to observe an eclipse in 1916. Another opportunity arose in 1918, when an eclipse took place in the U.S., but a series of unfortunate developments—including poor weather, not-yet-returned instruments that the University of California's Lick observatory had lent to the 1914 Russian expedition, and conflicts between group members—ended up with the results not being published.

Eddington helped enlist British Astronomer Royal Frank Dyson. Dyson was the first to propose an expedition to an upcoming total solar eclipse, on May 29, 1919. The path of the eclipse would run from North Brazil across the Atlantic, and pass Africa to the north. It would take place against a background of several bright stars. In November 1917, the Royal Society's Joint Permanent Eclipse Committee and the Royal Astronomical Society organized two expeditions to two different locations to take photographs. One, led by A. C. D. Crommelin and C. R. Davidson, was to Sobral in northeast Brazil, the other, which Eddington joined, was to Principe, an island about 10 miles long and 4 miles wide, owned by Portugal, about 120 miles off Africa's west coast. While the eclipse itself could be predicted, the weather could not be—and the weather at Sobral was especially worrisome because May was the last month of the rainy season.

The eclipse of May 29, 1919, was just another eclipse, yet another time that the earth entered into the cone-shaped shadow of its moon. But this one would turn out to be, scientifically, the most important eclipse in history.

In March 1919, the two expeditions left Greenwich, England, aboard the steamboat *Anselm*, and made a brief stop in Lisbon, Portugal, before arriving in Madeira. There the two groups split up. The Sobral group departed for Brazil on the *Anselm*, while the others stayed on at Madeira for 4 weeks to await a steamer to Principe.

When the Sobral expedition arrived at their a relatively barren location 80 miles inland, they set up their equipment in the racetrack of the local jockey club, after making sure that no races were planned before the eclipse. The Brazilian government supplied porters, bricklayers, carpenters, and interpreters, and an automobile—the first ever seen in Sobral—was brought all the way from Rio for the expedition's use. The team built a support structure to shelter their two telescopes against gusts. A minor calamity occurred when a whirlwind suddenly appeared and overturned the structure, but carpenters pillaged beams from elsewhere to fix it. A heavy rain fell on May 25, reminding the members of the expedition of the season. More disturbing was the discovery that the drive mechanism of the larger of the two telescopes was running unevenly, and that both instruments had focusing problems. On May 29, the moon's shadow began to sweep across the earth's surface. That morning, at Sobral, the members of the expedition awoke to an overcast sky. When the eclipse began, the sun was still behind clouds. But a minute before the eclipse became total, the sun emerged. For 6 minutes, Crommelin and Davidson took as many photographs as they could, exposing plates for 5 to 6 seconds each. "Eclipse splendid," they telegrammed.

The other expedition landed in Principe and set up shop at a plantation on the northwest side of the island. The regularly overcast sky made the team members apprehensive. On May 29, the team members awoke to a heavy thunderstorm, and cloud cover hung around the rest of the morning. When the eclipse started, the sun remained completely obscured. About half an hour before totality, the team members spotted glimpses of the now-crescent-shaped sun, raising their hopes. But the cloud cover never cleared completely. Eddington and his colleague, E. T. Cottingham, helpless, took pictures of the clouds and brief glimpses of the stars, hoping against hope that they might show something. "Through cloud. Hopeful," they telegrammed.

Over the next several months, groups set out computing the amount of deflection, taking into account the many sources of error.

"Joyous News Today"

In September, Einstein anxiously wrote to Lorentz to ask if he'd heard of the English results. On September 27, Lorentz telegrammed back:

> Eddington found star displacement at rim of sun, preliminary measurement between nine-tenths of a second and twice that value. Many greetings, Lorentz

Einstein promptly sent a postcard to his mother, Pauline, deathly ill and with only a few months to live:

> Dear Mother, joyous news today. H.A. Lorentz telegraphed that the English expeditions have actually demonstrated the deflection of light from the sun.[38]

Einstein also registered his excitement by sending a note to *Naturwissenschaften*.[39] But now that the anxiety was over, he could collect himself and grew more sangfroid. When his student Ilse Rosenthal-Schneider came to visit him, he showed her the telegram, saying, "Here, perhaps this will interest you." She later recalled,

> It was Eddington's cable with the results of measurement of the eclipse expedition. When I was giving expression to my joy that the results coincided with his calculations he said, quite unmoved, "But I knew that the theory is correct," and when I asked what if there had been no confirmation of his prediction, he countered: "Then I would have been sorry for the dear Lord—the theory *is* correct."[40]

On November 6, 1919, the joint meeting of the Royal Society of London and the Royal Astronomical Society was presided over by Sir Joseph J. Thomson, who had discovered the electron about a quarter-century earlier. He opened the meeting by saying, "I call on the Astronomer Royal to give us a statement of the result of the Eclipse Expedition of May last."[41]

Dyson:

> The purpose of the expedition was to determine whether any displacement is caused to a ray of light by the gravitational field of the Sun, and, if so, the amount of the displacement. Einstein's theory predicted a displacement varying inversely as the distance of the ray from the Sun's centre, amounting

to 1.75" for a star seen just grazing the Sun. His theory or law of gravitation had already explained the movement of the perihelion of Mercury—long an outstanding problem for dynamical astronomy—and it was desirable to apply a further test to it.

Any bending, Dyson continued, would have the effect of "throwing the star away from the Sun"; that is, making it look farther away. He briefly reviewed the events of the expedition, mentioning the defect discovered in the larger instrument. A good scientist, Dyson knew that all data are not created equal—some instruments perform better than others, in a way that one can evaluate independently of the result. Emphasizing the smaller instrument's measurements, Dyson concluded: "A very definite result has been obtained that light is deflected in accordance with Einstein's law of gravitation." Crommelin followed, offering a brief explanation of the defect in the larger instrument.

Eddington, a little outclassed by Dyson and by the quality of Crommelin's results, recounted his expedition and its frustrations with the weather. He painted his results as positively as possible, finding good in bad, pointing out that the cloud cover and uniform temperature at Principe had a beneficial side effect in reducing the mirror distortion that had adversely affected the larger Sobral mirror. Gamely admitting that he was "making the most of a small amount of material," Eddington noted that his result of 1.6" value for displacement at the limb "supports the figures obtained at Sobral." Disposing of the refracting matter explanation, Eddington concluded that the results support Einstein's law—the statement that light bends—though not necessarily Einstein's theory, the ideas about curvature of space behind it.

Thomson then took the floor again. He called the reports a "momentous communication." He said, "This is the most important result obtained in connection with the theory of gravitation since Newton's day, and"—alluding to the portrait of this early member hanging nearby—"it is fitting that it should be announced at a meeting of the Society so closely connected with him."

Some discussion lamented the mathematical complexity of the theory, which seemed out of the reach of most physicists. "I cannot believe," one said, "that a profound physical truth cannot be clothed in simpler language . . . Cannot Prof. Eddington translate his admirable treatise from the tensor notation into some such form?" A skeptic, citing the continuing absence of evidence of a spectroscopic shift, argued that the deflection result—"an *isolated* fact"—did not necessarily confirm Einstein's theory. "We owe it to that great man," he said, dramatically pointing to Newton's portrait—"to proceed very carefully in modifying or retouching his Law of Gravitation." But the arguments against Einstein's theory would all but die out within 2 years.

The experimental differences between his theory and those of Newton were tiny—small shifts in the positions of a few stars and spectral lines, and of a minuscule wobble in Mercury's orbit. But the differences could hardly be more profound, for they implied a fundamental difference in the way the universe was structured.

In Newton's theory, gravitation involves an attractive force—a tug—that each mass exerts on all others at a distance. That force operates instantly and everywhere, and is inversely proportional to the square of the distance between the masses. Masses experiencing that force respond by accelerating toward its source. Different masses are accelerated at the same rate, because the force pulls them in proportion to their mass: a small mass experiences a small pull, a greater mass a greater pull. In Einstein's theory, by contrast, gravitation involves in effect a curvature of space. That curvature is structured by the masses around it. When matter and energy move through space they follow the paths that are open to them.

It was one of the great rearrangements of fundamental concepts in the history of science. As Eddington wrote:

> The Newtonian framework, as was natural after 250 years, had been found too crude to accommodate the new observational

knowledge which was being acquired. In default of a better framework, it was still used, but definitions were strained to purposes for which they were never intended. We were in the position of a librarian whose books were still being arranged according to a subject scheme drawn up a hundred years ago, trying to find the right place for books on Hollywood, the Air Force, and detective novels.[42]

By 1921, the only prominent figure who continued to be disappointed in the theory was Einstein himself. To his exacting eyes, which wanted symmetry throughout the theory, the left-hand side was solid, for it was expressed in the geometry of space-time, while the right-hand side was not. He once compared it to a poorly planned building, one half of which was "fine marble," the other "low-grade wood."[43] Dissatisfied, he would spend much of the rest of his life in a futile effort to fix that building. Though he would work for over three decades on that repair, he would never succeed.

INTERLUDE

Science Critics

The equations of the gravitational field which relate the curve of space to the distribution of matter are already becoming common knowledge.

—Italo Calvino, *Cosmicomics*

Hence the pathetic paradox that Einstein's discoveries, the greatest triumph of reasoning mind on record, are accepted by most people on faith.

—*Time*, July 1, 1946

The process by which the public comes to understand new scientific developments often appears in the form of what might be called the Moses and Aaron model. Like Moses, the scientist seems to have one foot in the sphere of the divine, bringing into the human world some discovery from beyond. The meaning of this primordial activity is then communicated by some Aaron, who translates for the public via images and popular language.

The Moses and Aaron model is essentially a two-step process. We see it reenacted all the time—in the nightly TV news, for instance, when a spokesman or talking head tries to explain some novel development before the allotted 60 seconds run out. Some Aarons are more effective and entertaining than others.

Einstein's general theory of relativity poses a particular burden on would-be Aarons. A one- or two-step translation process is difficult to achieve. The theory involves complex mathematics, and unfamiliar ways of thought, that take physicists years to master. Learning the theory is like acclimatization into a culture, with no shortcuts. Sometimes—as *The New York Times* famously did after the November 6, 1919, announcement—journalists simply throw up their hands and say that a discovery cannot be explained to nonscientists. And physicist Hermann Bondi once said that members of the public would not understand relativity until they had relativistic toys to play with.

But talking about science to outsiders is like talking about a city to noninhabitants; what you say depends on the interests of your audience. If they intend to become inhabitants, you give them one kind of talk, focusing on regulations, institutions, laws, and so forth, that may take them awhile to master. If your listeners are just tourists with no intention to become inhabitants, on the other hand, you can focus on the public attractions, not go into too much detail, and safely condense a lot.

Many interesting attempts have been made to make general relativity accessible to a tourist audience. One way is by selecting and elaborating clever illustrations that show its implications in an accessible way: for instance, the twin paradox, wherein one twin who travels in space at near the speed of light ages differently from the other twin who remains on Earth; or the astronaut who cannot tell whether he or she is accelerating or in a gravitational field. Another way is through biographies: witness the phenomenal popularity of Walter Isaacson's recent book *Einstein: His Life and Universe*, or of Abraham Pais's earlier, more difficult but American Book Award–winning *"Subtle is the Lord": The Science and the Life of Albert Einstein*. Still another way to convey the meaning of general relativity to outsiders is by dramatic images, such as that of weights on a rubber sheet. A weight on a rubber sheet bends the sheet, by an amount that depends on the heaviness of the object, in a way that affects the path of marbles rolling past—in an analogous manner that an object distorts space, by an amount that depends on its mass, in a way that affects the paths of objects, including light, that pass through it. And some authors use all three methods at once, as Brian Greene in his wonderful books *The Fabric of the Cosmos* and *The Elegant Universe.*

Other clever strategies to convey complicated science to outsiders include Edwin A. Abbott's *Flatland: A Romance of Many Dimensions*, a famous novel involving an extended conversation between a square and a sphere that illustrates the problem of conceiving multiple dimensions. And Michael Frayn's brilliant play *Copenhagen* dramatizes an encounter between Niels Bohr and Werner Heisenberg that ends up illustrating many points about quantum physics.

What most needs bolstering in the contemporary discourse about science, however, is what might be called science critics. At least two individuals—the political scientist Langdon Winner and the philosopher Don Ihde—have called for such critics. In the arts, Winner points out, critics are instinctively understood as playing "a valuable, well established role, serving as a bridge between artists and audiences." A critic of literature, for instance, "examines a text, analyzing its features, evaluating its qualities, seeking a deeper appreciation that might be useful to other readers of the same text." Unfortunately, Winner lamented, the same kind of function is not performed in the sciences. One obstacle is that scientists tend to regard as suspect anyone who plays the role of critic, as if science critics are by definition objecting to science or insisting on its limitations.

Don Ihde, meanwhile, actively calls for science critics and even outlines what he thinks they should be like. "The science critic would have to be a well informed—indeed [a] much better than simply well informed—amateur, in [the sense of] a 'lover' of the subject matter, and yet not the total insider." The reason why the science critic must not be a total insider—just as an arts critic would not be a practicing artist or literary author—is because, as Ihde puts it, "we are probably worst at our own self-criticism."

Science critics, according to Winner and Ihde, would have an essential function. They would be there to assess the impact of science and technology on our political world (Winner) and on the human experience (Ihde). So, for example, Winner writes about the "politics" of technological artifacts, while Ihde writes about the transformation of experience by instruments. The kind of criticism advocated and practiced by Winner and Ihde, in short, judges the presence of science and technology in society, and has clear moral and political dimensions.

But there is another, complementary model for science criticism, one that involves another kind of interpretation, outlining the impact of scientific discoveries on our understanding of ourselves, the world, and our place in it. This model would require not a one-step translation process, but the kind of multiple roles that art criticism performs. It would involve a kind of "science criticism" just as elaborate and extensive as art criticism, whose presence is required for a thriving art culture. The necessary steps would include a complex field of several different niches of writings—books, articles, and columns, but also novels and plays, comments and reviews of these novels and plays, and so forth. This would allow the knowledge generated by science to have a cultural, and not merely an instrumental, presence, taking advantage of the processes by which culture enacts itself.

This model might be called *impedance matching.* In acoustics and electrical engineering, impedance matching involves taking a signal—produced inside a speaker, say—and putting it in a new environment with a different "load"—the surrounding environment—in a way that allows the signal to be heard. This is not a one- or two-step process, but requires a smooth and continuous matching or stepping down of the load. Scientific discourse, that is, bears one load—a heavy one—and public language a much different one. To connect the two effectively cannot be a matter of basic education plus popularization, but many different overlapping steps. Each of these steps requires more than rhetorical expertise, but connecting the signal with public issues and hopes.

Why should we bother? Our system seems to work relatively well as it is. Why make an effort to do more than paraphrase, to trace out the moral and spiritual impact of scientific work such as general relativity on the world? Part of the answer is to avoid being infantilizing and patronizing to the public. To the extent that Einstein's general theory of relativity represents humanity's best efforts at understanding the basic structure of the world, it is desirable for citizens—and not just professional scientists—to have the ability to acquire some *sense* of Einstein's general theory of relativity, some feel for what it means to our understanding of the universe, and a duty to make this possible. To put it more strongly, making this possible belongs to the human quest to acquire an understanding of ourselves and our place in the world. What is at stake is our own humanity.

CHAPTER NINE

"The Basic Equation of Quantum Theory": Schrödinger's Equation

$$\frac{d^2U}{dr^2}+\frac{2(a+1)}{r}\frac{dU}{dr}+\frac{2m}{K^2}\left(E+\frac{e^2}{r}\right)U=0$$

DESCRIPTION: How the quantum state of a system—interpreted, for instance, as the probability of a particle being detected at a certain location—evolves over time.
DISCOVERER: Erwin Schrödinger
DATE: 1926

The Schrödinger equation is the basic equation of quantum theory. The study of this equation plays an exceptionally important role in modern physics. From a mathematician's point of view the Schrödinger equation is as inexhaustible as mathematics itself.

—F. A. Berezin and M. A. Shubin, *The Schrödinger Equation*

The journey taken by the scientific community from Planck's introduction of the quantum to Schrödinger's assertion of its universal presence took barely a quarter-century.

When Planck introduced the idea in 1900, it was a tiny speck on the horizon. He used it to make classical theory work for black body radiation. The theory worked if we say that whatever absorbs and emits light (which he treated as "resonators") does so selectively—

only in integer multiples of a certain amount of energy. Many scientists saw this as a fudge, as problem avoidance rather than real science, and assumed that eventually they could discard the idea and it would drop back off the horizon.

Growing Extension of the Quantum

But in 1905, in a paper on the photoelectric effect, Einstein extended the idea. The quantum is not due to the selectivity of the resonators, he proposed, but to the fact that light itself is "grainy." By decade's end, the quantum had shown up in several different branches of physics. Many who had dismissed it now took notice.

In 1911, a landmark step was taken by Walther Nernst, a Prussian physical chemist who initially (like others) had dismissed quantum theory as the offspring of a "grotesque" formula, but who had used the theory to address what Thomson had called "Cloud No. 2," or the application of classical molecular theory of heat to experimental results involving low-temperature solids, gases, and metals. Nernst declared that, in the hands of Planck and Einstein (and, he should have mentioned, his own), the theory had proven "so fruitful" that "it is the duty of science to take it seriously and to subject it to careful investigations."[1] He organized a conference of leading scientists to do so, holding it in Brussels with the support of a wealthy Belgian industrialist named Ernest Solvay.

The conference, a milestone event, signaled that the quantum—the idea of a fundamental graininess to light and all other forms of energy—was in science to stay.

It was one of those events whose significance was immediately clear. Participants communicated the excitement to others who had not attended. Nobel laureate Ernest Rutherford, returning to Cambridge, England, described the discussions in "vivid" terms to a spellbound, 27-year-old Danish newcomer to his lab named Niels Bohr. In Paris, Henri Poincaré wrote that the quantum hypothesis appeared to involve "the greatest and most radical revolution in natural philosophy since the time of Newton."[2] Many scientists who were not present at the meeting caught its spirit from the proceedings. One was a 19-year-old Sorbonne student named Louis de Broglie, a recent convert to physics from an intended civil service career. De Broglie later wrote that the proceedings convinced him to devote "all my energies" to quantum theory.

But the quantum fit uneasily on the Newtonian horizon, even when it solved key problems. It was like a guest whom you could not get around inviting to an event, but who you also knew would be awkward and whose presence you would have to manage carefully. Consider what happened when Niels Bohr used it to explain Rutherford's until-then obscure idea about atomic structure. In 1911, Rutherford had proposed that atoms were like miniature solar systems, with a central core or "nucleus" surrounded by electrons. This contradicted a basic principle of classical mechanics: why didn't the orbiting electrons radiate away energy, as they should according to Maxwell's theory, and fall into the nucleus? Because, Bohr proposed, using the quantum idea, electrons could only absorb and emit radiation in specific amounts, and thus could only fit in a small number of stationary orbits or states inside the atom, able to absorb and emit only the energy required to jump between such states. It was an odd assumption indeed. It implied that atomic electrons—to employ an image that the American philosopher William James used to describe the stream of consciousness, which may have influenced Bohr—made "flights and perchings" amongst these states, without taking clear paths between them.[3] The states were what mattered, not the trajectories—whence the phrase, "quantum leap." Bohr applied this idea to the classic atomic test case—the hydrogen atom, a single electron orbiting a single proton. He showed how his assumption predicted the Balmer formula, an empirical formula for the spectral lines of hydrogen devised by a schoolteacher and numerologist.[4]

The flitting and perching things—now applied only to light, but soon to matter—would soon create a classification problem. In the classical horizon, the tiniest things came in two basic types: particles and waves. Particles were discrete things: each had its own definite position and momentum, and always followed a specific path in space and time. Waves were continuous things: they spread out spherically from their source without specific position or direction, smoothly broadening and thinning in space and time. Scientists used different theories to describe particles and waves. Particles were addressed by Newtonian theories that assumed masses were located at specific points and pushed by forces and had a definite momentum and position at every moment. Waves were addressed by Maxwellian theories that used continuous functions to describe how processes smoothly evolve in space and time. Both theories were well developed and deterministic: you input information about the initial state, turn the crank, and out popped a prediction of future behavior.

In which bin should the flitting and perching things be placed? They seemed to have aspects of each. How was that possible?

Einstein provided some of the answer in his 1905 photoelectric effect paper. Traditional optics, he said, treats light as waves because it involves light in large amounts and averaged over time. But when light interacts with matter, as when it is emitted and absorbed, it does so on very short timescales, when it may well be grainy, localized in space, and with energies in integer multiples of $h\nu$ ("quanta" of light later called "photons"). This idea, he proudly wrote to a friend, was "very revolutionary."[5]

For the next 20 years, physicists tended to be partisans of either particle theory or wave theory, trying to extend one or the other to cover quantum phenomena.

Einstein carried the theoretical banner for the particle side, though not without some reluctance. In an important paper of 1916, he extended his idea that light is absorbed and emitted in the form of physically real quanta, each having a particular direction and momentum (a multiple of $h\nu/c$), and—making a general if somewhat overstated point—proclaimed that "radiation in the form of spherical waves does not exist."[6] This process conserved energy, he now showed, for the amount emitted at one end equaled the amount absorbed at the other. But Einstein also found that he had to incorporate statistics in his theory to make it work, in the form of "probability coefficients" that described the emission and absorption of quanta.[7] He found this to be a painful sacrifice, but hoped it would be temporary, expecting that his work would soon be replaced by some deeper understanding. Einstein's experimental allies included Arthur H. Compton, who in 1923 demonstrated the "Compton effect," that when photons bounce off electrons they both come from, and rebound in, definite directions.[8]

One champion of waves was physicist Charles G. Darwin, grandson of his more famous naturalist namesake—though he, too, was somewhat dissatisfied with his role. Darwin believed light was emitted as waves, but realized that accommodating quantum phenomena such as the photoelectric effect would severely tax wave theory. In 1919, he wrote a "Critique of the Foundations of Physics," in which he foresaw fundamental changes ahead. Quantum phenomena, he prophesied, might force physicists to abandon long-cherished principles. They might have to entertain wild ideas, he wrote—tongue-in-cheek—such as to "endow electrons with free will."[9] The least wild thing, he finally decided, would be to keep wave theory by abandoning the conservation of energy for individual events, having it conserved only on average.

Darwin found a sympathizer in Niels Bohr. In 1924, Bohr enlisted two others—Hendrik Kramers and John Slater—in an attempt to eradicate Einstein's radical idea and develop a more conventional approach that used wave theory to account for how light is emitted and absorbed, and for the photoelectric and Compton effects.[10] The authors found they had to pay a heavy price to murder Einstein's idea; they would indeed have to abandon the conservation of energy, conserving it only on average, along with any hope of having a visualizable picture of the mechanics of how light is emitted and absorbed.

The word "visualizable"—*anschaulich*, for the Germans—became something of a technical term in physics around this time. For something in a theory to be visualizable or intuitable two things had to happen: the variables in the theory had to be connected with physical things like mass, position, energy, etc.; and the operations in the theory had to be connected with familiar operations, such as point-by-point movement, action at a distance, and so forth. Thus for something to be visualizable, or *anschaulich*, it did not necessarily have to be Newtonian, because something strange and non-Newtonian can still be visualized, as long as it unfolds in space and time. If something were *anschaulich* it merely meant that a flip-book-like description could be created in which the pages were like slices of time, locating where everything in an event is at every moment—and that when you ruffle the pages, what is on each page blends smoothly into what is on the next.

But the sacrifices of the Bohr-Kramers-Slater theory—abandonment of the conservation of energy and of *anschaulichkeit*, were regarded as too extreme not only by most physicists but even by at least one of its authors; Slater later claimed to have been coerced into signing his name. Few were surprised when, less than a year after publication, the Bohr-Kramers-Slater proposal was decisively refuted by experiment.

The Bohr-Kramers-Slater paper is a unique document in the history of science. It is renowned among historians for being both obviously wrong and strongly influential. It was strongly influential because it brought to a head the conflict between particle and wave theory. It said: *This* is the kind of sacrifice you have to pay in order to keep what you have. The partisans of each side were only being cautious and conservative, trying to preserve those elements of classical theory which they thought most robust. But quantum phenomena were resisting.

At the end of its first quarter-century, indeed, quantum theory was a mess. Historian Max Jammer called it "a lamentable hodgepodge of hypotheses, principles, theorems, and computational recipes rather than a logical consistent theory." Each problem had to be first solved as if it were a classical situation, then filtered through a "mysterious sieve" in which quantum conditions were applied, weeding out forbidden states and leaving only a few permissible ones. This process involved not systematic deduction but "skillful guessing and intuition" which resembled "special craftsmanship or even artistic technique."[11] A theory was needed that gave the right states from the start. To put it another way, quantum theory was more like a set of instructions for coming up with a way to get from point A to point B, when what you really wanted was a map.

Then, in 1925, came two dramatic breakthroughs from two very different people: Werner Heisenberg and Erwin Schrödinger. Each, struggling to act conservatively by sacrificing as little of the classical framework as possible, ended up a revolutionary.

Heisenberg, who at age twenty-four was young even by physics standards, tried to save classical mechanics by abandoning it at Nature's bottom rung. Inside the atom, he declared, not only do particles and electron orbits have no meaning, but neither do even such basic classical properties as position, momentum, velocity, and space and time. And because our imaginations require a space-time container, this atomic world cannot be pictured. We have to base our theories, he said, on what he called "quantum-theoretical quantities" that are unvisualizable, or *unanschaulich*. The next chapter outlines the steps Heisenberg took in developing his approach. At one point, Heisenberg noticed an odd feature: certain sets of quantum-theoretical quantities were noncommutative under the peculiar definition of "multiplication" they obeyed: the order in which they were multiplied mattered. He initially found this feature awkward, and tried to ignore it—but soon came to embrace the feature as the keystone of quantum mechanics. In 1925 he wrote "On the Quantum-Mechanical Reinterpretation of Kinematic and Mechanical Relations," which provided a method for calculating quantum states that lacked both particles and waves. It utilized mathematical methods that we call matrices to provide a formal, mathematical apparatus into which one plugged experimental data, turned the mathematical crank, and out popped the allowed states. His supervisor, Max Born, quickly saw that Heisenberg had rediscovered matrices. But matrix mechanics, as it was called, was difficult to use, and many physicists resisted a theory that told them they could not picture Nature's bottom rung.

Erwin Schrödinger (1867–1961)

Schrödinger, then thirty-eight, was ancient by physics standards. His approach covered much the same territory, but used a familiar tool of classical mechanics: a wave equation that he had developed, with continuous functions that described processes unfolding smoothly in space and time. For Schrödinger, the bottom rung was made of something quite *anschaulich*: waves.

Enter Schrödinger

Erwin Schrödinger had arrived at the University of Zürich in 1921.[12] New professors were asked to deliver a formal talk to a general audience, and Schrödinger's was called "What Is a Natural Law?" In it, he endorsed the possibility that "the laws of nature without exception have a statistical character."[13] True, Maxwell had introduced statistical laws into physics to describe the behavior of systems, such as gases, consisting of large quantities of small things. But such laws were conveniences—approximations, cheats—used because in practice our knowledge is limited. In principle, we could track the behavior of each and every molecule to predict the system's behavior; plug numbers for the forces and masses into Newton's laws, turn the crank, and out would pop predictions of past and future behavior. And while Einstein had used probabilities in his 1916 paper, he thought these would be temporary. Though Schrödinger could not know it, the work he was about to do would shortly be interpreted as implanting statistics permanently into nature's laws, without a deeper underlying law.

Illness slowed Schrödinger's work for a few years, but by 1925 he was studying aspects of quantum theory and participating in joint colloquia between the university and the nearby technical school, the Eidgenössische Technische Hochschule (ETH). One day in the fall of 1925, one of the ETH's organizers, the Dutch physicist Pieter Debye, asked Schrödinger to report on a recently published thesis by Louis de Broglie, now a newly graduated French physicist, who had indeed thrown himself into quantum theory after reading the proceedings of the Solvay conference. He introduced the notion of a wave process accompanying electrons, using Planck's rule $E = h\nu$ to connect the momenta of electrons with a wavelength. With this assumption, he was able to explain the quantization conditions of the old quantum theory. And so at one of the next ETH colloquia, Schrödinger dutifully explained the young Frenchman's idea that the right orbits were obtained if one assumed electrons had integer wavelengths.

Debye, sitting in the front row as was traditional for someone of his eminence, dismissed the idea as "rather childish." If something were a wave, he said, it needed a proper wave equation.

What he meant seems to have been this: waves usually refer to something that is waving. Elsewhere in physics, waves are solutions of equations of motion for that "something." De Broglie had identified a wave associated with an electron, but he gave no clue as to what was waving or what its equation of motion should be.

Schrödinger, unlike most people at the ETH colloquium, took Debye's remark seriously—and was also stirred by Einstein's remark about de Broglie's work that an "undulatory field is connected with every motion."[14] Schrödinger finished a paper he was writing on the quantum theory of gases, and left on a skiing vacation in Arosa with an old girlfriend whose identity is a mystery, for his diary for 1925 is lost and obvious suspects have been ruled out. A colleague once commented that Schrödinger "did his great work during a late erotic outburst in his life"—though the remark is puzzling, for while thirty-eight may be late for physics, it is not for eros. Schrödinger's biographer writes that, "like the dark lady who inspired Shakespeare's sonnets, the lady of Arosa may remain forever mysterious," and that "whoever may have been his inspiration, the increase in Erwin's powers was dramatic . . . he began a twelve-month period of sustained creativity activity that is without a parallel in the history of science."[15]

On December 27, Schrödinger wrote to Wien from Arosa:

> At the moment I am struggling with a new atomic theory. If only I only knew more mathematics! I am very optimistic about this thing and expect that if I can only . . . solve it, it will be very beautiful.[16]

Schrödinger returned from Arosa to Zürich on January 9, seemingly still struggling. But shortly thereafter, he opened another colloquium talk with the words, "My colleague Debye suggested that one should have a wave equation; well, I have found one!" And in a remarkable series of six papers published in 1926—"Quantization as a Problem of Proper Values," published in four parts and "undoubtedly one of the most influential contributions ever made in the history of science,"[17] plus a paper on the transition between the quantum and the classical world, and another on the relation between wave mechanics and matrix mechanics—Schrödinger presented his wave equation and examined its implications.

Schrödinger's equation incorporated a wave function that he called "a new, unknown ψ," and related (as de Broglie had) the wavelength to a momentum, and the frequency to an energy. The behavior of the atomic realm, Schrödinger was proposing, is made up of waves of the ψ field—which was something like a charge density, a kind of particle fog, he initially felt—waves that added up, interfered, created nodes, and so forth. This "eminently visualizable" picture, Schrödinger claimed in the first part of his multipart paper, allows us to picture such experimentally observable things as "how two colliding atoms or molecules rebound from one another, or how an electron or α-particle is diverted, when it is shot through an atom."[18] It gives rise, he said, to a picture of electron states in atoms as in effect standing waves—waves that, as in a violin string, retain their basic shape while oscillating. And, he remarked in the second part of the paper, he hoped to be able to demonstrate that his theory would show how wave groups or "packets" form with "relatively small dimensions in every direction" that "obey the same laws of motion as a single image point of the mechanical system;" that is, that act like single particles.[19]

It would not be that simple. The ψ field, after all, was a mathematical quantity whose properties would determine physical, observed properties only when supplemented with additional steps. When Schrödinger put in these steps, he found that they included the complex number *i*. He was initially disturbed by its presence in his wave theory and tried to get rid of it. He could not. He was disturbed because a complex number involves two components—a real part and an imaginary part—and its presence implied that the wave function had a phase that was not directly observable. A phase is like a clock, a cycling phenomenon, and the fact that the phase had an imaginary part meant that it had an aspect that could not be directly measured. It was oscillating in time in a way that could not be seen from the outside, from "reality." Schrödinger's equation described something "waving" in a multidimensional or "configuration" space.

What, then, was this ψ? In his first paper, Schrödinger wrote that originally he had hoped "to connect the function ψ with some *vibration process* in the atom, which would more nearly approach reality than the electronic orbits, the real existence of which is being very much questioned to-day." This vibrational process, that is, would create something like what physicists were calling a quantum state, a discrete, discontinuous thing built up out of continuous processes. "[T]o imagine that at a quantum transition the energy changes over from one form of vibration to another," he commented at the end of his first paper, is far more satisfying than "to think of a jumping electron," for "the changing of the vibration form can take place continuously in space and time," and he contrasted his work with the notable failure of the wave theory of Bohr, Kramers, and Slater. Though for the moment, he continued, he would not pursue these thoughts and satisfy himself with presenting his ideas in "neutral mathematical form."[20]

Furthermore, he hoped that superpositions of waves could create a "wave packet" that kept itself together—like a traveling wave in a pond with a stable, sharply defined peak—which would explain what was happening when this ψ field acted like a single particle.

Schrödinger soon discovered that the straightforward "wave-packet" intuition of this kind would not be possible. In the fourth and final part of his "Quantization" paper series, he wrote, "the ψ-function itself cannot and may not be interpreted directly in terms of three-dimensional space." It was a wave only in a strange formal or "configuration space." Still, Schrödinger's approach did what he wanted: it talked about the atomic world in terms of our world—space, time, waves, and so forth using equations with which physicists were familiar and knew how to use with relative ease. The picture he provided was intuitive, more or less. And interpreting the equation—relating it to our more commonsense notions of the world—was important, if only to guard against views that would give up on rational calculation by asserting that God or something supernatural were pushing the particle around.

But others soon wrested Schrödinger's intuitive interpretation away from him.

Interpreting the Waves

In the summer of 1926, Göttingen physicist Max Born—who was Heisenberg's supervisor, and a developer of matrix mechanics—published his work on atomic collisions, as between electrons and atoms. Collisions, after all, were the central focus of classical physics, and Born regarded it as one of the key issues in understanding the atomic realm. He had struggled with the matrix approach, in vain, and had reached a startling conclusion. "[O]nly Schrödinger's formalism proved itself appropriate for this purpose," Born declared; "for this reason I am inclined to regard it as the most profound formulation of the quantum laws."[21] But he also had unwelcome news for Schrödinger. Born could not make any sense out of the claim that the ψ-function referred to the electron's charge density. Schrödinger's equation, Born concluded, does not tell us information about the state of an event, but rather about the *probability* of a state. The ψ function that Schrödinger's equation described as continuously moving through space, interfering and interacting with potentials, was not some substantial field but probabilities. "We free forces of their classical duty of determining directly the motion of particles and allow them instead to determine the probability of states."[22] Born thus fashioned a strange hybrid with elements of both wave and matrix mechanics: it incorporated both continuity and causality on the one hand, and discontinuity and probability on the other. "The motion of particles conforms to the laws of probability, but the probability itself is propagated in accordance with the law of causality."[23]

A few months later, another interpretive step was taken by Wolfgang Pauli, another former assistant of Born's. While Born had interpreted ψ as about the probabilities of states, Pauli now said it was about the probability of particles—that ψ^2 represented the probability of an electron being at a particular position. This was still further from Schrödinger's interpretation of his function, for it completely stripped the ψ function of reality. The ψ function was about the possibility, not the actuality, of something—of the click of a counter, of the presence of a particle. The actuality had to be brought about by setting up equipment and having some subatomic performance be enacted—an interaction between whatever it is described by the wave function and the world.

The Born-Pauli interpretation, a marriage of particle and wave theory, quickly became the interpretation adopted by most physicists. But the marriage had a strange cost, for features of each were lost. When Newton's laws governed particles, the particles were observable and the laws were deterministic—one plugged in the initial states of the particles and turned the crank to get predictions. The same was true for Maxwell's laws governing waves: waves were observable things, with fully measurable properties, and Maxwell's laws were deterministic, describing how they governed over time. Both particle and wave theory, that is, were about predictables and observables. The Born-Pauli interpretation now married particle and wave theory together—but a part of each was destroyed. The Schrödinger wave waves in configuration space. The particles were observable but lost their predictability; the waves were predictable but lost their observability. Observing the position and momentum of something does not allow you to make predictions about where you will see it next. The wave is used to predict the probability of another event, but after the event is observed, the wave has no more value and has to be discarded or "reset," modified to incorporate new information.

Nowadays, this interpretation is often presented in a misleading way. Instead of saying that the wave function is discarded or reset when a measurement is made, one often hears that the wave function "collapses." The imagery captures the idea that, before the event happened—before you detected the particle, say—it could be anywhere, so one thinks that the event or particle is everywhere. The image this conjures up is of a structure, extended through space, suddenly getting sucked up at a single point. It's a vivid but deceptive image. The wave is just a probability, not a "thing." (It is the one merit of what is called the pilot wave idea that nothing collapses; the wave merely deposits the particle already in it.) The wave function, whose purpose is only to give probabilities, has flowed along—predictably, deterministically—but once an event happens the function has exhausted its purpose, and must be reset.

Refashioning John Wheeler's famous remark about Einstein's theory of general relativity, Stony Brook physicist Alfred S. Goldhaber says the following about Schrödinger's equation: "The wave tells the particle where to go, the particle tells the wave where to start and stop."

It is a great irony that this approach, devised to be intuitive, often suffers thanks to a false picture.

The Schrödinger equation implied a radical change in the events on the world-stage. No longer could it be assumed that, eventually, you could plug in numbers, turn the crank, and get predictions. Instead, you plugged in numbers, turned the crank, and got . . . probabilities. You got the likelihood of an event occurring in a particular place. And nothing could help you more than that: not more information, not adding more pieces to the machinery. No flip book could be constructed, for if you replayed an event over and over, a particle would wind up in different spots on the page, its location a question of averages.

Schrödinger himself never warmed to this interpretation, calling it a "resignation."[24] It is "convenient," he wrote, but we cannot allow ourselves to "get off so easily." We should keep trying to work out the interpretation, the causal mechanism, he insisted. He pointed to some seeming implications that he found "quite ridiculous," including the image of the now-famous cat locked in a box along with a diabolical device that kills the cat upon the decay of a radioactive nucleus. Since the decay is defined by ψ, it appears that whether the cat is alive or dead is so defined as well, seemingly leading to the conclusion that the cat's existence, too, is superimposed, and is half-alive and half-dead. While the conclusion is false, the image is a brilliant demonstration of the flaws of extending the theories of the microworld to those of the macroworld.

But while Schrödinger's wave methods are used by all the workers in the field, his premises about the wave structure of reality have been ignored; the interpretation worked out by Heisenberg's Göttingen allies came to be the favored one. "Schrödinger's methods proved indispensable, writes historian Mara Beller. "His philosophy did not."[25]

Born's interpretation of Schrödinger's equation changes the notion of what a complete theory of the world consists of. Conventionally, we expect a complete theory to tell us something about reality. Most theories that physicists teach and use fall short by deliberately cutting corners. These theories provide not a complete picture, but a model that is an idealized version of any real situation. For example, the ideal-gas law ignores well understood things like van der Waals forces and hard-core repulsion, but we do not mind because talking about an ideal, rather than a real, object buys us something: a tremendous ease of application. This is what one of my colleagues calls a "harmlessly fudging theory." Such theories are harmless because these limitations do not threaten normal assumptions about the world. We are cutting corners, know that we are cutting corners, and know that the absence of the corners we are cutting does not affect our impression of the world.

Yet Schrödinger's equation, as interpreted by Born, is different. It makes us aware that how we interact with the world affects what we bring into it. It makes us explicitly aware, that is, of ourselves as interacting with the world in doing something like taking a measurement. The idea that we are agents comes explicitly into view. We are not watching something happen on a stage and then measuring it, we are making the stage even as we do so. The prevailing interpretation of Schrödinger's equation brings this to the fore in a way classical physics does not. It is our own expectations that determine whether we find this a sacrifice or an advance. The giant leap forward of quantum mechanics does require a substantial rethinking of what it means to "understand" nature, and how to characterize "reality," in a way that many scientists at the time experienced as what Heisenberg called a sacrifice—so painful a one, in fact, that they struggled mightily against having to make it. Many still do.

INTERLUDE

The Double Consciousness of Scientists

. . . a world which yields him no true self-consciousness, but only lets him see himself through the revelation of the other world. It is a peculiar sensation, this double-consciousness, this sense of always looking at one's self through the eyes of others, of measuring one's soul by the tape of a world that looks on in amused contempt and pity.

—W.E.B. DuBois

Schrödinger's work was greeted with not just skepticism but hostility from Heisenberg and other supporters of matrix mechanics, who revealed what Mara Beller called "a dogmatic preference for older conceptions rather than a dispassionate objectivity," and who were all the more annoyed because of the beauty and simplicity of Schrödinger's approach compared to the ungainly and complex matrix methods.[1] Schrödinger, meanwhile, made no secret of his scorn for matrix mechanics. The formation of quantum mechanics, indeed, was a remarkable episode in the history of science in the way hostile emotions flared in private, and even formal published papers are marked by an "unusually emotional undertone."[2] Envy, rivalry, anger, disbelief, conviction, stress, hope, despair, dejection—all can be found in the documents. Yet the existence of this emotional undercurrent to scientific research is omitted from most historical accounts.

Most histories of science run something like this: An unexpected discovery is made. Explanations are applied, but none work well. New equipment is built to make new measurements, but the explanation is still incomplete. The phenomenon is looked at from yet another side, with other instruments and measurements. And so on.

This is what might be called the "standard model" of how science works. It emphasizes the collective and impersonal dimension, and downplays the experiences of specific individuals. The principal structural ingredients are discoveries, instruments, measurements, and theories. It is allied with a conception of science as a way of eradicating mysteries and controlling nature. Its familiarity often leads scientists to tell their own stories in ways that emphasize these ingredients, reinforcing the standard model. In this model, it's as if the emotional life and experiences, personal successes and disappointments, and so forth, of the person comprised a train running down one track, and the scientific career—the research program, discoveries, events, and so forth—running down another track. And the two tracks are entirely separate, driven by different kinds of locomotives—two sides of one person.

But listen carefully to the scientists speaking about their work—as some popular biographies do—and you can hear another story as well, highlighting human experience. In it, motivating forces include excitement at the discovery, puzzlement at why the explanations do not work, curiosity about what might explain it, growing perplexity as more explanations do not work, imagination at devising new instruments, and wonder as the explanations shed a different kind of light than thought at the outset, even awe at learning something fundamental. As these more fully integrated pictures of science in process show, the standard model has limits, there is something beyond it, and it is ultimately destined to be superceded. It is due to be succeeded by a program of grand unification, in which these two tracks are seen to be merged, in which science is done by individuals, not dividuals, whose life and work are part and parcel of the same person.

To see what I mean, look at the collection of Richard Feynman's letters published a few years ago. In them, you see Feynman's character, poses and all, as inextricably intertwined with his craft. You see that his curiosity, presumption, haranguing, and desire to set people straight were seamlessly interwoven—that the physicist and educator and his character cannot be disentangled. The same force fueled both his scientific inquiries and his interactions with others. "The real fun of life," he wrote, "is this perpetual testing to realize how far out you can go with any potentialities." And you see this testing in the way he dealt with cranks, editors, ordinary people, and with nature. This provides a taste, I think, of what lies beyond the standard model.

We can also see this grand unification in Einstein. Here science historian Gerald Holton has written a good article called "Einstein's Third Paradise." Einstein's "first paradise" refers to the intensely religious phase Einstein went through in childhood—a period of "the religious paradise of youth," he called it. This phase is well attested to by Einstein himself, and by his sister Maja. This paradise ended when he was about twelve, after reading popular science books that revealed to him that not all the biblical stories could be true. He also discovered the joys of Euclidean plane geometry after being given a little book on it—he called the book "holy," and a "wonder." He became acquainted with other works of science that presented nature as "a great, eternal riddle," contemplation of which could give him "inner freedom and security." He called this, too, a "Paradise." Breaking away from the first paradise to enter the second paradise, he wrote, was an attempt to "free myself from the chains of the 'merely personal,' from an existence dominated by wishes, hopes, and primitive feelings."

Biographers have tended to contrast these two paradises, deeming them two separate, disconnected phases of his life, from a religious to a nonreligious phase. But Holton disagrees. At the heart of Einstein's mature identity Holton sees a fusion of the first and second paradises, "where the meaning of a life of brilliant scientific activity drew on the remnants of his fervent first feelings of youthful religiosity."

In this third paradise, Einstein seems to exemplify someone who had feelings that we can call religious and that were essential to his work but who did not credit the existence of a Master Mechanic. As he says in one letter, he was a "deeply religious unbeliever." This third paradise, then, is the kind of thing that would be described in what I called the grand unification. Consider Einstein's speech honoring the sixtieth birthday of Max Planck. In it, Einstein said the search for a simplified, lucid image of the world was not only a scientific goal, but corresponded to a deep psychological need. A scientist could make the effort to pursue this goal the "center of gravity of his emotional life." And, Einstein added, pursuing the most difficult scientific problems requires "a state of feeling similar to that of a religious person or a lover." Holton then mentions instances by Einstein and others in which they were brought to great despair, or great joy, by developments in science, in which the psychological commitment of these people cannot be treated as separate from the tasks they set for themselves. The science and the personal commitment are bound up with each other. Holton treats Einstein's drive to unify apparently different phenomena as an example of this interpenetration of emotional life and career. He points to a letter to Grossmann in 1901, referring to his very first paper, on capillarity, which unifies opposing behaviors of bodies. "It is a wonderful feeling," Einstein wrote (echoing Kant), "to recognize the unity of a complex of appearances which, in direct sense experiences, appear to be quite separate things." And in another letter, 15 years later, Einstein says that he is "driven by my need to generalize." Holton points out that, practically, too, "Einstein lived under the compulsion to unify." He loathed nationalisms, and dreamed of a unified world government. Holton sums up: "No boundaries, no barriers: none in life, as there are none in nature. Einstein's life and his work were so mutually resonant that we recognize both to have been carried on together in the service of one grand project—the fusion into one coherency." Likewise, Holton says, "there were no boundaries or barriers between Einstein's scientific and religious feelings." In his writings on science and religion late in life, Einstein often uses the same phrases to refer to the aims of science and religion. "I maintain that the cosmic religious feeling is the strongest and noblest motive for scientific research. . . . A contemporary has said not unjustly that in this materialistic age of ours the serious scientific workers are the only profoundly religious people." And again, "The most beautiful experience we can have is the mysterious. A knowledge of the existence of something we cannot penetrate, our perceptions of the profoundest reason and the most radiant beauty, which only in their most primitive forms are accessible to our minds—it is this knowledge and this emotion that constitute true religiosity; in this sense, and in this alone, I am a deeply religious man." Thus in Einstein, too, we can see glimpses of what lies beyond the standard model: an account of science in which character and personal feeling are not marginal to the scientific process, not a prelude to a person's scientific labors, but what sustains them and carries them forward.

CHAPTER TEN

Living with Uncertainty:

The Heisenberg Uncertainty Principle

$$\Delta x \Delta p \geq \frac{\hbar}{2}$$

DESCRIPTION: Establishing the position of a particle in a small region of space makes its momentum uncertain, and vice versa, and the overall uncertainty is greater than or equal to a certain amount.
DISCOVERER: Werner Heisenberg
DATE: 1927

Everyone understands uncertainty. Or thinks he does.

—Werner Heisenberg character, in
Michael Frayn's play *Copenhagen*

We owe many a debt to Werner Heisenberg. As one of the founders of quantum mechanics, he left a huge legacy to physics. As the inventor of the uncertainty principle, he also left a huge legacy outside physics. Albert Einstein may be more widely recognized by the public—and his theory of relativity often crops up in popular culture—but Heisenberg has had a similar far-reaching impact on public discourse and popular culture.

While most nonscientists may recognize Einstein's equation $E = mc^2$, generally they are aware that its effects are noticeable only in certain restricted conditions, and that its meaning is truly clear only to physicists. The same is not true of Heisenberg's equation $\Delta x \Delta p \geq \hbar/2$, the Heisenberg uncertainty principle, which seems to have a spiritual meaning to the public that is at once profound and transparent. Browse through any bookshop's new-age section, for instance, and you'll find wild claims confidently asserted about the uncertainty principle, such as that its implications are "psychedelic" and that it heralds "cultural revolution." Strange interpretations turn up even in academic circles. Consider the following conversation, published in *American Theater*, between the well-known theater director Anne Bogart and Kristin Linklater, the noted vocal coach:[1]

> **Linklater**: Some thinker has said that the greatest spiritual level is insecurity.
> **Bogart**: Heisenberg proved that. Mathematically.
> **Linklater**: There you are.

But where are we, exactly? And how did we get here? The uncertainty principle sprang from a purely mathematical approach to atomic physics, where it has a well-defined and highly restricted scope of applicability.

The Path to Helgoland

Werner Heisenberg, son of a professor of Greek at the University of Munich, had the character one often associates with poets: dashing good looks, a physical frailty including severe vulnerability to allergies, excellent musicianship, and a sensitive and often emotional responsiveness to the world around him.[2] He also had a sharp but imaginative intellect and a willingness to risk unconventional but mathematically rigorous means to fit theories to experimental data. And he had the terrific fortune to be reared amid one of the sharpest and most ruthlessly demanding scientific communities ever, whose members included Niels Bohr, Max Born, Pascual Jordan, Hendrik Kramers, and Wolfgang Pauli.

These theorists were largely distributed among three centers of research: Munich, Göttingen, and Copenhagen. Each had a distinctive character. Munich was experimentally oriented, Göttingen was a world-renowned center of formal mathematics, and Copenhagen had a rigorous philosophical approach to the quantum world stemming from its founder and leader, Bohr. The intense and often brutally frank exchanges of this community of physicists—carried on in personal conversations as well as in letters, drafts of work in progress, and copies of published papers—kept anyone who dared participate to a high standard. Many times a thought initiated by one person was completed by another. Heisenberg, a central player, circulated among the three centers, and his insights, too, often arose in conversation.

Werner Heisenberg (1901–1976)

In July 1923, Heisenberg completed his doctoral exam at Munich, and had arranged to work under Max Born in Göttingen that fall. But Heisenberg, a supposed wunderkind, had nearly failed thanks to his almost total ignorance of experimental physics—he could not even explain how a storage battery worked—and passed only after aggressive intervention by one of his examiners. The day after the humiliating exam he showed up at Born's door in Göttingen, unannounced and despondent, to confess to Born the embarrassing news and ask if Born still wanted him. Born was supportive, and Heisenberg left, reassured, for a summer trip he regularly took with a youth group.

This was just the time in quantum theory that historian Max Jammer described as an unruly mess, when those problems that could be solved at all were first analyzed classically, then restricted by quantum conditions to obtain a few "allowed" states of motion. Heisenberg, still only 21 years old, was determined to make it all rational.

He knew that classical physics had to be the starting point. "The concepts for quantum mechanics can only be explained by already knowing the Newtonian concepts," he remarked much later. "That is, quantum theory is based upon the existence of classical physics. This is the point that Bohr emphasized so strongly—that we cannot talk about quantum physics without already having classical physics."[3]

In classical physics, all events take place inside a four-dimensional space-time stage or playing field. Everything is at a specific place at any and every time. When things move from one place to another, they do so in response to definite forces and take definite paths. Classical physics mainly concerns itself with what happens when things are disturbed, and tracks which forces produce which effects. The path of each thing can be followed—and predicted—like that of a smoothly flowing stream, with the thing moving continuously and smoothly from each point to the next. Physical properties, which can all be measured, propagate smoothly and continuously through space-time in a mechanical way. Classical physics thus provides a confident ontology, or vision of what the ultimate elements of the universe are and how they interact. This kind of event is therefore *anschaulich*.

But a quarter-century of attempts to devise classical models of quantum phenomena had failed. Stimulated by the debates swirling in his rich intellectual environment, young Heisenberg began to wonder if *that* were not the problem; if the effort to construct pictures of the world inside the atom—the positions and paths of electrons, and the dimensions and frequencies of their orbits—was doomed from the start. He had heard Pauli remark that models of atomic events had "only a symbolic sense" and were classical "analogues" of the quantum phenomena.[4] Wasn't the lesson of Maxwell's path to his equations that sometimes one had to abandon mechanical explanations to capture reality? Wasn't it likely, then, that when theorists construct models based on what experimenters measure, these models are only symbols of a reality that humans cannot picture?[5] Progress in science usually involves sacrifice, Heisenberg once wrote, a sacrifice that is at the cost of our claim to understand nature. What had to go, this time, Heisenberg and his colleagues thought, might be visualizability.

Heisenberg therefore decided to make a virtue of necessity by junking the attempt to produce theories that picture how atomic events unfold on a space-time stage. Drawing on the appreciation for formal structures he had acquired at Göttingen, he would seek a purely mathematical description of what experimenters actually observed: the frequencies and amplitudes of the light emitted by electrons. These descriptions would need to respect only the correspondence principle—that large quantum numbers obeyed classical laws—and certain other constraints such as the conservation of energy. But there would be no need to have measurable properties or continuously propagating functions; indeed, discontinuity seemed to Heisenberg the principal distinctive feature of the quantum realm, and thus would characterize its theory.

This insight was momentous. It has been likened to Copernicus's insight into the structure of the solar system. Both changed the viewpoint from which scientists had been used to regard the world, treating what had been naïvely assumed to be the image of objective reality as the more complex product of an interaction between the human observer and nature.

The step was revolutionary, but the way had been prepared not entirely by Heisenberg. First, he was utilizing theoretical tools acquired from Bohr, Born, and others, and abandoning the space-time stage only because that seemed the price one had to pay to use them. Second, Heisenberg had a fine precedent in the strategy Einstein used in 1905 to give birth to special relativity. Einstein had abandoned the traditional meaning of "simultaneous" as "happening at the same space-time instant," and redefined it in terms of what an observer could see. Heisenberg hoped to achieve a similar breakthrough by abandoning the traditional conception of "position" and "momentum" inside the atom—which were unobserved but inferred quantities—and redefine these in terms of what experimenters saw from the outside: the frequencies and amplitudes of spectral lines. Finally, it was not such a radical step to give up trying to construct a theory that tried to picture what could not be observed given the utter failure of all the theories that had tried.

But like most revolutions, it had long-range consequences that would take years to become clear. If to be a "thing" meant occupying a specific place at a specific time, this approach meant "eliminating the concept of a particle, or 'thinghood,' from the atomic domain."[6] This approach essentially replaced the Newtonian ontology of nature, in which its most fundamental pieces are all objectively present in a particular place at a particular time, with a new ontology involving, as one philosopher of science put it much later, "a subtle subjectivity at the very heart of the scientific enterprise."[7] The subjectivity relates to the fact that our pictures of the atomic world are not an image of objective reality but are partly a function of the human mind constructing the pictures. The subtlety related to the fact that it was not yet clear what role the mind played.

But so much was not at all clear at the time. Heisenberg's path unfolded in fits and starts over the first few months of 1925. He cowrote a paper with Hendrik Kramers in Göttingen with equations containing no classical variables but only frequencies and amplitudes. Kramers's contribution was an important clue, for he showed that only when these frequencies and amplitudes are associated with pairs of states does one get the correct matrices. Then Heisenberg injured himself skiing and spent several weeks in Munich recovering. He visited Copenhagen and Göttingen, took another trip to the mountains, and by the end of April returned to Born's institute in Göttingen to prepare to teach a summer session. All of these visits prepared Heisenberg to try to rewrite Bohr's quantum descriptions of electron momentum (p) and position (q) in purely mathematical terms. He did not tell his supervisor what he was up to, keeping the idea, as Born once put it, "dark and mysterious."[8]

Then, Heisenberg once recalled, "My work along these lines was advanced rather than retarded by an unfortunate personal setback."[9] In May, he was hit by an attack of hayfever so severe that he asked Born for 2 weeks off. Born agreed, and Heisenberg headed for Helgoland, an isolated, rocky island in the North Sea that is inhospitable to grass, weeds, and other allergen producers. The evening before his departure, the landlady of the *Gasthaus* who showed him his room was so horrified by his swollen face that she assumed he had been in a fight. On the island, once able to take up his work again, he tried to see if his ideas were consistent with the conservation of energy. When they were, he grew excited. He made mathematical errors, owing to his condition and fatigue, but caught them and continued working late into the night, finally sorting out everything by about 3 A.M.

> At first I was deeply alarmed. I had the feeling that, through the surface of atomic phenomena, I was looking at a strangely beautiful interior, and felt almost giddy at the thought that I now had to probe this wealth of mathematical structures nature had so generously spread out before me. I was far too excited to sleep, and so, as a new day dawned, I made for the southern tip of the island, where I had been longing to climb a rock jutting out into the sea. I now did so without too much trouble, and waited for the sun to rise.[10]

Heisenberg returned to Göttingen late in June, and was soon scheduled to leave to lecture in Cambridge. In a few days he dashed off a paper, "On the Quantum-Mechanical Reinterpretation of Kinematic and Mechanical Relations."[11] The word "reinterpretation" (*Umdeutung*) reveals Heisenberg's audacity: it was the manifesto of a new approach to atomic physics. The abstract boldly declared that the aim was "to establish a basis for theoretical quantum mechanics founded exclusively upon relationships between quantities which in principle are observable." We have not been able to "associate an electron with a point in space" based on the experimental information, he continued, and "in this situation it seems sensible to discard all hope of observing hitherto unobservable quantities, such as the position and period of the electron." Heisenberg was groping in the dark here; it would turn out that quantum mechanics would contain the possibility of measuring position and momentum to any degree of accuracy, just not simultaneously. The paper showed how to compile tables of amplitudes and frequencies associated with transitions between states—he called such tables "quantum-theoretical quantities"—and how the tables could be related by a new kind of calculus, which he called "quantum-mechanical relations."

The paper associated physical quantities with tables whose rows and columns were labeled with the "allowed" quantum states that Bohr had postulated in his groundbreaking paper on the hydrogen spectrum. This had been done before (for example, Einstein's *A* and *B* coefficients are "tables" labeled with two states), but Heisenberg applied the idea to a more fundamental set of quantities and went a step further and found a rule for "multiplying" two such tables to make formulas similar to those used in classical mechanics. This was new, and it opened the way to do quantum calculations far beyond the limited capacity of the previous attempts, such as Born's, at a "quantum mechanics."

Heisenberg then hit a snag. The tables and the multiplication rule he invented for them obeyed a new kind of algebra that mathematicians had discovered long before, but was unfamiliar to most physicists, himself included. Most strikingly, the rule did not follow the "commutative law," the mathematical principle according to which the order in which one multiplies two numbers does not affect the result: $ab = ba$. When Heisenberg used his new calculus to multiply one quantum-theoretical table (let's call it A) by another (B), the result depended on the order: $AB \neq BA$. The feature "was very disagreeable to me," he said later, and try as he might he could not rid his theory of it.[12] "I felt this was the only point of difficulty in the whole scheme, otherwise I would be perfectly happy." Heisenberg then did what many people do when a nuisance threatens to spoil a promising new invention: he swept it under the rug. He waved at it in a single sentence—"Whereas in classical theory [*AB*] is always equal to [*BA*], this is not necessarily the case in quantum theory"—mentioned circumstances in which the difficulty does not arise, and then dropped the subject. Heisenberg concluded his paper with a disclaimer of the sort that is often seen in early papers in a field, wondering whether it was "satisfactory" or "still too crude an approach" to quantum mechanics. The answer, he declared, would have to await "a penetrating mathematical investigation."[13]

After he finished, around July 9, Heisenberg gave Born a copy of the paper, asking his supervisor to see whether it was worth publishing, and to see whether he could investigate the basic idea, which Heisenberg knew seemed awkward and even bizarre. Born so promised—but set the paper aside for a few days, exhausted after a semester teaching and from research he was doing with his other assistant, Pascual Jordan.

Born read the paper only after Heisenberg's departure. Impressed, he sent it on to the *Zeitschrift für Physik*, and on July 15 wrote to Einstein that Heisenberg's work appeared "very mysterious, but is surely correct and profound."[14] But Born also had a nagging feeling about Heisenberg's tables and the strange mathematical rules used to multiply them. It looked so familiar! After a restless week in which he could hardly sleep, it hit Born that he had seen this peculiar structure in his high school math classes. His intrepid young assistant had reinvented the wheel. The tables were what mathematicians called matrices, arrangements of numbers (or variables) in rows and columns—though Heisenberg's tables had *infinite* numbers of elements. And Heisenberg's funny quantum-mechanical relations were actually the most natural way that mathematicians had discovered to "multiply" matrices.

Born was overjoyed. The mathematics of matrices gave him a framework in which to investigate and systematize Heisenberg's work. He knew that matrices can be noncommutative—the order in which one multiplied them mattered. This explained Heisenberg's embarrassing difficulty that, for instance, the matrix $\mathbf{p}$ associated with momentum and $\mathbf{q}$ with position did not commute; the matrix $\mathbf{pq}$ was not the same as $\mathbf{qp}$ (by convention, physicists often indicate matrices with bold symbols). But there was more. This pair of variables—known as canonically conjugate variables—was not commutative, but in a special way. Though Born could not prove it, the difference between $\mathbf{pq}$ and $\mathbf{qp}$ seemed to be a specific matrix proportional to Planck's constant: $\mathbf{pq} - \mathbf{qp} = \mathbf{I}h/2\pi i$, where $\mathbf{I}$ is the unit matrix—"ones" along the diagonal entries and zeros everywhere else. Born wrote later, "I was as excited by this result as a sailor would be who, after a long voyage, sees from afar, the longed-for land, and I felt regret that Heisenberg was not there."[15]

A few days later, on July 19, Born ran into Pauli on a train, excitedly explained how Heisenberg's paper could be translated into matrix language, and asked his former assistant if he wanted to collaborate in investigating the topic. Pauli was dismissive, and sarcastically accused Born of trying to "spoil Heisenberg's physical ideas" with "futile mathematics" and "tedious and complicated formalism." (Historians find this remark humorous, for Heisenberg's ideas in this case were formal and even more tedious than conventional matrix analysis.) The next day, July 20, Born approached Jordan, who happened to be knowledgeable in the mathematics of matrices. Within a few days the two were able to show how to derive the relation $\mathbf{pq} - \mathbf{qp} = \mathbf{I}h/2\pi i$ from Heisenberg's work, and again Born was awestruck: "I shall never forget the thrill I experienced when I succeeded in condensing Heisenberg's ideas on quantum conditions in the mysterious equation $\mathbf{pq} - \mathbf{qp} = \mathbf{I}h/2\pi i$, which is the center of the new mechanics and was later found to imply the uncertainty relations."[16]

By the end of September they sent off a paper, "On Quantum Mechanics." It carried out the "penetrating mathematical investigation" that Heisenberg had hoped for, and was the first formulation of what became known as matrix mechanics. The math was unfamiliar—many physicists had to bone up on matrices to understand the paper—and its methods were unwieldy, but it worked on the limited number of problems for which calculations could be carried through to completion. The authors sent a copy to Heisenberg, who by then had left Cambridge and was in Copenhagen. He showed the paper to Bohr, saying, "Here, I got a paper from Born, which I cannot understand at all. It is full of matrices, and I hardly know what they are."[17] But after Heisenberg brushed up on matrices, he too, shared their excitement, and on September 18 wrote to Pauli that Born's bright idea, $\mathbf{pq} - \mathbf{qp} = \mathbf{I}h/2\pi i$, was the foundation of the new mechanics. Heisenberg, Born, and Jordan began a feverish discussion by letter, and Heisenberg interrupted his stay in Copenhagen and returned to Göttingen so that the three of them could finish work on another paper generalizing the results of the Born-Jordan paper before Born departed for a long-scheduled trip to the U.S. in October.[18]

The result was a paper written by Born, Heisenberg, and Jordan entitled "On Quantum Mechanics II," known to historians of physics as "the three-man paper." Its central feature is what they called the "fundamental quantum-mechanical relation," the strange equation $\mathbf{pq} - \mathbf{qp} = \mathbf{I}h/2\pi i$. The paper is a landmark in the history of physics, for it is the first map of the quantum domain. Around the same time, Pauli published a paper in which he had successfully—though with considerable difficulty—applied matrix mechanics to the test case of the hydrogen atom.

Yet few besides its creators fully recognized the importance of matrix mechanics, for the appreciation of its value was hampered by several obstacles. One was its complexity: while matrix mathematics was not intrinsically that difficult, Heisenberg's application of it appeared to be horrendously complicated, and most physicists had to take matrix mechanics on faith while they struggled to master it. Typical was the reaction of George Uhlenbeck, then a student at the University of Leiden, who remarked much later, "Everything became these infinite numbers of equations that you had then to solve, and so nobody knew exactly how to do it."[19] Others were put off by the *unanschaulichkeit*—by the fact that matrix mechanics deliberately refrained from providing a picture of the mechanics of the atomic domain, and that its fundamental terms, the matrices, were strictly speaking meaningless, just formal symbolic artifacts.[20] Still others were bothered by the failure to explain the transition from the microworld to the macroworld—between the unvisualizable world without space and time to the familiar space-time container that we and our imaginations live in. Many scientists therefore took, as historian Mara Beller once wrote, a "wait and see" attitude, and even its originators viewed it as but a first and imperfect step toward an adequate theory.[21]

But shortly after the three-man paper appeared in February 1926, its authors had unwelcome company.

Matrix vs. Wave Mechanics

Schrödinger's first and second papers on wave mechanics appeared in the *Annalen der Physik* in March and April 1926. Wave mechanics mapped the same terrain as matrix mechanics, but physicists found the map much easier to read. It lacked the obstacles of matrix mechanics. First, the math was part of the bread-and-butter training of classical physicists, who had been using and solving wave equations since their high school days. Second, wave equations were visualizable. Physicists saw water, sound, and light waves, and their properties—frequency, amplitude, and wavelength—smoothly and continuously propagating around them daily. They had trained themselves to see other wave properties such as nodes and interference. There was the small matter of the ψ-function, which existed in multidimensional "configuration space"—three dimensions for each particle in the system—but even that seemed somewhat visualizable, as something that traveled through space or stayed "perched" in a standing-wave-like way when bound inside an atom. Third, wave mechanics provided a natural way of describing the transition from the microworld to the macroworld, as Schrödinger's third paper that year showed, as particle-like groupings or packets of waves moved along the classical paths, which were the rays perpendicular to the phase fronts of the ψ-function.[22]

Small wonder most physicists took to wave mechanics. Planck was awestruck, Einstein ecstatic. U.S. physicist Karl Darrow reported that wave mechanics "captivated the world of physics" as it promised "a fulfillment of that long-baffled and insuppressible desire" to return to classical physics, its comfortable and continuously propagating functions.[23] A flood of papers used Schrödinger's approach to tackle atomic issues. Allies of the Göttingen physicists reacted badly: Heisenberg called it "too good to be true," Dirac reacted with "hostility," Pauli called it "crazy."[24] But many of them soon fell under its spell. Pauli, who had just laboriously worked out the theory of the hydrogen spectrum using matrix mechanics, found wave mechanics much easier to use for the same purpose. Born wrote to Schrödinger that he grew so excited upon reading the first wave mechanics paper that he wanted "to defect . . . to continuum physics . . . [to] the crisp, clear conceptual foundations of classical physics,"[25] though his ardor soon cooled.

At first the conflict played out in arguments about the scientific merit of the two approaches: Which approach did the better job? The hydrogen atom—which Pauli solved with both methods—was a first key test case. It was the drosophila fly or lab rat of atomic physicists, the problem that any model had to tackle first—for the hydrogen atom had been successfully analyzed with the old quantum theory, in close agreement with experiment, meaning that formula could be compared. Another test case was to account for the transition between the quantum and the classical world, or how to get from its rungs to ours. Schrödinger had shown that wave mechanics had an answer to this, but none was apparent yet for matrix mechanics. Yet another key problem was how to handle collisions between things in the atomic world, which would require showing how a system evolved over time.

The answer to the question of which approach had more scientific merit was soon resolved: In May, in his fourth paper of 1926, Schrödinger proved that, mathematically speaking, the two approaches were identical.[26] Pauli reached the same conclusion. It was not yet clear how to handle all the test cases, but the demonstration of mathematical equivalence showed that neither approach had more or less mathematical merit than the other. Scientifically, though, wave mechanics could do more than matrix mechanics, for it was essential for analyzing the continuous part of the spectrum.

As Heisenberg's biographer David Cassidy notes, however, this conclusion only restructured the conflict so it could begin in earnest. With the issue of mathematical equivalence settled, the partisans now were liberated to argue about the physical interpretations of the theories. These were dramatically opposed: wave mechanics—at least as interpreted somewhat hopefully by Schrödinger—portrayed the atomic world as woven out of continuous processes that are causally responsible for what seem to be discontinuous events, and as unfolding in space and time, while matrix mechanics portrayed the atomic realm as lacking continuous processes and causal relations, as not related to space and time, and as not being a world at all in any way humans can imagine. This conflict was less decidable and more emotional than the one about scientific merit, for it reflected the adversaries' sense of what physics was all about, of what the world was, and of the most fundamental relationship between human beings and the world.

Yet, as Beller notes, "in the initial stages of the controversy over interpretation *nobody* had a clearly articulated position, let alone a handle on the 'truth.'"[27] The restructured clash now forced the partisans to argue for the physical interpretations of their theories. Schrödinger had to argue that, at its most fundamental level, the world was full of continuities, and that to describe it one did not require Heisenberg's awkward formal methods. Schrödinger also had to explain how wave packets could hold together, elaborate the meaning of the ψ-function, and demonstrate how the discontinuities of quantum phenomena arise from continuous wave processes.

Schrödinger, finally, had to admit the problem of a wave that existed in multidimensional configuration space. Heisenberg and his allies had to argue that the world was full of discontinuities, and that it was misleading to present it otherwise. They had to provide some way of connecting the formal symbolic terms of matrix mechanics to familiar properties, and show why, to the extent that wave mechanics was visualizable, it was false. As partisans are wont, they were not always consistent; Beller has shown how each side "cheated" a little, incorporating aspects of the other in order to make the theories work. But this new and ferocious clash now set the stage for the emergence of both the uncertainty principle and the Copenhagen interpretation of its meaning.[28]

Schrödinger began sniping already in his paper proving the identity of the two approaches. While they were indeed identical, he said, he was "discouraged, if not repelled" by the "very difficult" mathematical methods of matrix mechanics and by its lack of visualizability.[29] Later he said that it was "extraordinarily difficult" to attack atomic issues such as the transition problem so long as one has to "repress intuition" and "operate only with such abstract ideas as transition probabilities, energy levels, etc."[30] He wrote to Wien that the avowals about the necessity of restricting physics to observables "only glosses over our inability to guess the right pictures."[31]

Heisenberg's language was at least as sharp; he described wave mechanics as "disgusting" and as "garbage." To the extent that wave mechanics was visualizable it was false, he claimed; physicists who use matrix mechanics are less deluded and thus see deeper into nature.[32] As Born once said, "Mathematics knows better than our intuition."[33]

The conflict was soon fought in face-to-face encounters. In July 1926, Schrödinger and Heisenberg met for the first time at a conference in Munich, where Schrödinger had many supporters. Schrödinger gave two talks about wave mechanics, and Heisenberg stood up at the end to object that no theory relying on continuous processes could possibly explain the discontinuities of quantum phenomena, such as Planck's radiation law and the Compton effect. The audience appeared to be on Schrödinger's side. Heisenberg seems to have influenced no one, and left feeling defeated. He went to Copenhagen, where he stayed for several months working with Bohr. The two disagreed—Bohr argued that we *must* use classical concepts to describe experiments, with Heisenberg disagreeing—but they honed their arguments why quantum discontinuities implied that space and time could not be defined, meaning that the quantum realm could neither be represented by theories involving continuous functions nor imagined by the human mind whose picturing ability requires a space-time container.

The next bout between Schrödinger and the matrix allies took place 3 months later, in October 1926. Bohr invited Schrödinger to visit Copenhagen, which was largely matrix territory (though the Copenhagen group had already begun to use some wave mechanics as a tool); Schrödinger, intellectually honest and seemingly riding the popular side, was happy to visit the opposition's headquarters. But he was utterly unprepared for what followed. Bohr met Schrödinger at the train station, almost immediately began pressing his case, and continued arguing day and night for several days. Bohr had arranged for Schrödinger to stay at his house, so that every possible minute could be used. As Heisenberg recalled:

> Bohr was an unusually considerate and obliging person, but in this kind of discussion, which concerned epistemological problems which he thought were of vital importance, he was capable of insisting—with a fanatic terrifying relentlessness—on complete clarity in all argument. Despite hours of struggle, he refused to give up until Schrödinger had admitted his interpretation was not enough, and could not even explain Planck's law. Perhaps from the strain, Schrödinger got sick after a few days and had to stay in bed in Bohr's home. Even here it was hard to push Bohr away from Schrödinger's bedside: again and again, he would say, "But Schrödinger, you've got to at

least admit that. . . ." Once Schrödinger exploded in a kind of desperation, "If you have to have these damn quantum jumps then I wish I'd never started working on atomic theory!"[34]

With Heisenberg at his side, Bohr persuaded Schrödinger into making a (temporary) retraction. But it did not last, and the originator of wave mechanics was soon back writing papers about it. In November 1926, indeed, he assembled his six seminal papers on wave mechanics—the four part series "Quantization as a Problem of Proper Values" that appeared in the *Annalen der Physik*, plus his papers on the boundary problem and on the identity between wave and matrix mechanics—and had them published as a book.

By this time, Born had contributed his novel interpretation of wave mechanics. Trying to understand collisions between an electron and an atom, Born had carefully examined Schrödinger's claim that the ψ-function referred to the electron's charge density, found it did not make sense, and concluded that it does not tell us about the state of an event but rather about its *probability*. Pauli then wrote his letter to Heisenberg in which he proposed that ψ^2 represented the probability, not of states, but of particles at particular positions. This amounted to a partial restoration of the space-time container and of visualizability. It did not entail that the orbits or paths of electrons from one place to another could be visualized, but that, however they got there, they did *have* positions.[35] The classical properties do exist, and can be measured precisely. Still, it involved the bizarre notion that the strange function that Schrödinger said flowed through space was not a real thing but the probability that a real thing could be found at that spot. At the time, the philosophical novelty of this was not noticed. "We were so accustomed to making statistical considerations," Born remarked later, that "to shift it one layer deeper seemed to us not so very important."[36]

In the same letter of October 19 in which Pauli made his proposal about the interpretation of the wave function, he also noted implications for the vexing $\mathbf{pq} - \mathbf{qp}$ issue. Heisenberg had been arguing that neither of the conjugate variables—the noncommuting terms—referred to classical variables such as positions or momenta that could be measured with precision together. Pauli was now saying that one of the pair could be—but if so, the other was only known as a probability. This made the noncommutativity even stranger. "The physics of this is unclear to me from top to bottom," Pauli told Heisenberg. "My first question is: why can only the p's, and not *simultaneously* both the p's and the q's, be described with any degree of precision?" He was baffled. "You can look at the world with p-eyes or with q-eyes, but open both eyes together and you go wrong."[37] What could this mean?

Heisenberg's response was delayed because he had a hard time retrieving Pauli's letter from his excited Copenhagen colleagues who were sharing it. Heisenberg finally sent a reply on October 28. He still did not buy the implied restoration of visualizability and classical variables, dismissing Born's "rather dogmatic" view as "only one of several possible interpretations." The $\mathbf{pq} - \mathbf{qp} = \mathbf{I}h/2\pi i$ relation, he continued to insist, meant that individual **p**s and **q**s were meaningless. "Above all, I hope there will eventually be a solution of the following type (but don't spread this around): That time and space are really only statistical concepts, something like, for instance, temperature, pressure, and so on, in a gas. It's my opinion that spatial and temporal concepts are meaningless when speaking of a single particle, and that the more particles there are, the more meaning these concepts acquire. I often try to push this further, but so far with no success."

And a few weeks later, on November 15, Heisenberg presented to Pauli what seemed a conclusive argument why the discontinuities of the quantum world made the very concept of individual **p**s and **q**s meaningless.[38] Let's say an object such as an electron is at a specific point. Its velocity is defined in terms of the rate at which it moves continuously through points vanishingly close to it—but if space-time is discontinuous, and electrons flit from one state to another, it must lack velocity by definition! A week later, Heisenberg returned obsessively to the issue.[39] Because the world is discontinuous, the "c-numbers" (classical numbers) imply that we know way too much about what is happening. "What the word 'wave' or 'corpuscle' mean, one does not know any more."

Pascual Jordan now stepped in to challenge Heisenberg. In effect, Jordan played contrarian to Heisenberg's postulate-of-impotence claim that single electrons could not have positions and momenta. What was stopping experimenters from measuring them? Observing equipment is made of atoms, and atoms rattle about at room temperature due to their thermal motion, imposing a practical limit on accuracy. So what if we somehow set up the equipment to make a measurement at absolute zero where thermal motion stops; or, in what amounts to the same thing, what if we use highly energetic probes such as α particles, whose rattling is negligible and whose paths can be tracked?

Born and Pauli had considered the theoretical possibility of fixing one conjugate variable, and noted that the other could only be said to have a certain probability. Jordan was now pointing out experimental conditions in which physicists could indeed measure what was supposedly forbidden: the "probability of finding an electron in a certain place." It's not unobservable theoretically, just difficult experimentally.

Jordan's article troubled Heisenberg.[40] The day after it appeared, on February 5, 1927, Heisenberg wrote to Pauli that he found Jordan's paper "nice enough but not very exact in places," because he still thought phrases like "probability of finding an electron in a certain place" were conceptually meaningless. But if things such as the time and position of individual electrons made experimental sense, they had to make theoretical sense. If they made theoretical sense, his approach had to be wrong.

In all these discussions, there was never any question that the mathematics was correct. It was the interpretation that was at issue, and even the nature of interpretation. Bohr demanded more of interpretation than Heisenberg, and both demanded more than Schrödinger.

Heisenberg was still in Copenhagen, working at Bohr's institute and living in the garret apartment of Bohr's brother Harald. After supper, Bohr would come by, pipe in hand, and the two would argue about the state of quantum mechanics until the morning hours. The demanding conversation was beginning to wear on their relationship, and the two grew testy. Realizing this, Bohr left to go skiing. One evening while Bohr was gone, Heisenberg took a walk in Faelled Park, behind Bohr's institute. He pondered **p**s and **q**s in theory and in experiment. He thought about Jordan's microscope. He was as convinced as ever that something *had* to be wrong with Jordan's example. Jordan had brought Heisenberg down to earth, derailing his single-minded focus on theoretical meaning, for it forced him to stop philosophizing about concepts and think operationally about what experimenters did. Suppose you looked at a particle at absolute zero; this would mean bouncing a photon off it and capturing the photon in the instrument's lens. But that would disturb the electron's position. If you wanted to avoid this, you would have to use a less energetic photon. But the longer the wavelength of the photon, the less precisely you knew its position! The problem might occur, Heisenberg excitedly realized, because of the interaction between the instrument and what you are seeking to measure—between the tools you were using to observe and the system observed.

Dawn of a New Era

Heisenberg then did what he often did when excited: he wrote a letter to Pauli. This one, dated February 23, was unusually long—fourteen pages. The shift in his thinking inaugurated by Jordan's article is evident at the outset, for he describes several thought experiments involving measurements of **p**s and **q**s. Then he writes, "One will always find that all thought experiments have this property: When a quantity p is pinned down to within an accuracy characterized by the average error p, then . . . q can only be given at the same time to within an accuracy characterized by the average error $q_1 \approx h/p_1$."

This is the uncertainty principle. Like many other equations, its first appearance was not in the form in which it is now famous. Today it is usually written as an inequality: $\mathbf{\Delta x \Delta p} \geq \hbar/2$.

The uncertainty principle was a conceptual breakthrough. While Born, Pauli, and Jordan had considered cases where one conjugate variable was exactly determined and the other a probability, Heisenberg now showed these are limiting cases, and in between is a spectrum of other cases where neither value is exact. A margin of uncertainty is unavoidable. If the uncertainty (Δx) in the position of, say, an electron is small, then the uncertainty in the momentum (Δp) must be large enough to keep the product, $\Delta x \times \Delta p$, on the order of h. If the position of an electron is measured with such precision that the uncertainty is very small, the corresponding uncertainty in the momentum becomes very large. And Heisenberg told Pauli that this was a direct consequence of $\mathbf{pq} - \mathbf{qp} = \mathbf{I}h/2\pi i$, whose interpretation finally seemed clear. Heisenberg put particles back in a space-time stage, at least for the moment, but gave them decidedly unclassical properties.

Heisenberg quickly wrote a paper bearing his thoughts, "The Visualizable [*anschaulich*] Content of Quantum Kinematics and Mechanics." It set out to explain to classically trained physicists how quantum mechanics might be visualized in classical terms, and to do so redefines the word in the first sentence: "We believe we understand the visualizable (anschaulich) content of a theory when we can see its qualitative experimental consequences in all simple cases and when at the same time we have checked that the application of the theory never contains inner contradictions." This definition is too quick and convenient, designed so that Heisenberg eventually can make his theory fit it. But never mind; Heisenberg then says that it might seem difficult for quantum mechanics to fit this definition, for whenever $\mathbf{pq} - \mathbf{qp} = \mathbf{I}h/2\pi i$ holds, it is unclear what we mean by things like position and velocity, and we need to clarify matters by specifying experimental conditions. So let's say we observe an electron under a microscope that illuminates it with light. Because it is very small, we have to use energetic light: γ-rays. But if we use energetic light on tiny things, the Compton effect comes into play; the photon collides with our little electron, and abruptly and discontinuously shoves it away. Heisenberg wrote:

> This change is the greater the smaller the wavelength of the light employed—that is, the more exact the determination of the position. At the instant at which the position of the electron is known, its momentum therefore can be known [only] up to magnitudes which correspond to that discontinuous change. Thus, the more precisely the position is determined, the less precisely the momentum is known, and conversely. In this circumstance we see a direct physical interpretation of the equation $\mathbf{pq} - \mathbf{qp} = -ih$.

Heisenberg was cavalier about his use of the unit matrix **I** in this equation, and it is frequently omitted in the literature as well. He continued by quantifying this interpretation:

> Let q_1 be the precision with which the value q is known (q_1 is, say, the mean error of q), therefore here the wavelength of the light. Let p_1 be the precision with which the value p is determinable; that is, here, the discontinuous change of p in the Compton effect. Then, according to the elementary laws of the Compton effect p_1 and q_1 stand in the relation
>
> $$p_1 q_1 \sim h$$

Now comes an odd thing whose significance has not been noted until recently, by John H. Marburger, III. Heisenberg proceeded to say that this equation is "a straightforward mathematical consequence of the rule equation $\mathbf{pq} - \mathbf{qp} = -ih$," *but he does not show it.* There is no derivation of the uncertainty relation in Heisenberg's paper! While it was accepted by Heisenberg and Bohr, and it was clearly a good conjecture, neither bothered to prove it, and the first proof of the principle to which Bohr refers is flawed.[41]

This "visualizable" paper was less radical than the "reinterpretation" paper of 2 years before. It did not argue that an electron lacked position or velocity, only that it had no *simultaneous* definite position and velocity, leaving the door open for one or the other to have a precise value. Heisenberg restored enough visualizability to claim that "quantum mechanics should no longer be considered as abstract and non-visualizable." In a kind of coup de grâce, he quoted Schrödinger's remark about how "disgusting and frightening" matrix mechanics is, to set up a retort that the real enemy is Schrödinger's misconceived understanding of visualizability. The atomic world is visualizable, but what one could visualize was clearly not classical. A careful reading leaves one unsure whether Heisenberg was really committed to visualization at all. As Beller writes, "Heisenberg assumed the classical picture of the world in order to refute it."[42]

After finishing the paper, Heisenberg wrote to Jordan that he felt "very, very happy" that after a year of being continuously suspended, he now felt the "discontinuous ground under my feet."[43] And Pauli was thrilled. "He said something like, 'Morgenröte einer Neuzeit' "—the dawn of a new era.[44]

But the new era got off to a rocky start. When Bohr returned and Heisenberg showed him the paper, Bohr spotted several blatant errors. Even in the atomic world, Bohr pointed out, energy and momentum are conserved, and if you disturb an electron by knocking it with a photon you can still figure out its momentum by catching the photon, eradicating the uncertainty. Yet, Bohr continued, Heisenberg's idea was still correct, but because of the wave nature of particles. You cannot determine the momentum of recoiling particles precisely—not even if you use electrons instead of photons—because they all spread out in a wavelike manner just as Schrödinger's equation described, which is why you use a microscope lens to focus them. But this meant acknowledging that Schrödinger's waves played an essential role in the theory. The conversation quickly deteriorated, and neither Bohr nor Heisenberg budged from his deeply entrenched position: Bohr said you needed waves, Heisenberg that you could do without them. Bohr told Heisenberg not to publish the paper, and the latter eventually burst into tears with frustration.[45] But as Beller points out, these tears are as much due to Bohr's ruthlessness as to Heisenberg's stubbornness.

Heisenberg ignored Bohr's advice and refused to withdraw or even fix the paper; he merely appended a brief note, entitled "Addition in Proof," which stated that "Bohr has brought to my attention that I have overlooked essential points in the course of several discussions in this paper." But he did not fix the overlooked points.

For months, Bohr and Heisenberg continued to disagree about the interpretation of quantum mechanics. Both agreed that the mathematics was right and, as Einstein noted, had to guide the interpretation. But Bohr had a better idea of how to go about it. Heisenberg argued that you could use either matrix or wave language, Bohr that you needed both. Heisenberg's position was essentially Platonist: he wanted to say that the mathematics alone describes what exists in the atomic realm. Bohr's position was Kantian: nature forces human beings to experience and imagine according to certain (classical) categories and schemata structured by a space-time stage; as Marburger puts it, reality is a macroscopic phenomenon. These categories and schemata are adequate for macroscopic events, and appropriate for the classical physics which sought to provide the theory for such events. But these categories and schemata do not apply to microscopic events—to apply them and assume they are valid is to make what might be called the *macroscopic fallacy*. Still, we cannot get around these classical schemata in our thinking and imagining. Therefore, Bohr concluded, in our thinking about the microscopic world we are forced to depend on classical categories and schemata—such as position and momentum—but these categories are to be used in overlapping, nonclassical ways, as in "complementary" pairs. We have to abandon the notion that the concepts and schemata adequate for sensible phenomena in the macroscopic world correspond to what is real in the microworld. Bohr's Kantian approach therefore severed an ontological connection between the quantum theory and the world of "real" phenomena. Down there, it's stranger than we can say. "[A]n independent reality in the ordinary physical sense can neither be ascribed to the phenomena nor to the agencies of observations."[46]

Late in 1927, however, Bohr was scheduled to take a trip to the U.S., and he and Heisenberg were anxious to finalize an interpretation before his departure, so the two agreed to a truce, mainly on Bohr's terms. The truce was made public that September, at a celebration in Lake Como of the hundredth anniversary of Alessandro Volta's death. Bohr gave a speech proposing an awkward accommodation of wave and matrix mechanics, and Heisenberg stood up at the end to signal his approval. Waves and particles, Bohr said in effect, are ways we speak about events in the atomic realm. Neither way is entirely accurate, but the two ways have overlapping but restricted spheres of application. They are, Bohr declared, complementary ways of speaking about something of which we can have no direct knowledge. As he once put it, "There is no quantum world. There is only an abstract physical description. It is wrong to think that the task of physics is to find out how nature is. Physics concerns what we can say about nature."[47]

Thus the origin of what has become known as the Copenhagen interpretation of quantum mechanics. It was not universally appreciated. Einstein called it "shaky," adding that "The Heisenberg-Bohr tranquilizing philosophy—or religion?—is so delicately contrived that, for the time being it provides a gentle pillow for the true believer from which he cannot very easily be aroused."[48] In complete theory, he wrote years later in 1935, with Boris Podolsky and Nathan Rosen, "every element of the physical reality must have a counterpart in the physical theory." Einstein tried to argue, unsuccessfully, that the incompleteness of quantum mechanics was a flaw revealing that there had to be more to be discovered, so-called hidden variables, the discovery of which will make its formulations refer directly to the real world. He pressed the argument for years, with Bohr countering that position and momentum were inherently classical concepts, inapplicable to events in the microworld except in loose and, strictly speaking, inaccurate ways.

The Copenhagen interpretation—that somewhere beyond or beneath the macroscopic world lurks something that we cannot visualize, and that is made visualizable by an ensemble or arrangement of things whose behavior is macroscopic—amounts to a clear, logical interpretation, and appears to be the simplest one consistent with all experimental and theoretical constraints. It is an interpretation that makes us all uncomfortable, but that is a psychological phenomenon, not an argument for or against the interpretation.

INTERLUDE

The Yogi and the Quantum

The idea of intermediate kinds of reality was just the price one had to pay.

—Werner Heisenberg

In 1929, 2 years after the appearance of the uncertainty principle, a physicist at Harvard University named Percy Bridgman—a future Nobel laureate—published an article in *Harper's Magazine* about the meaning of the uncertainty principle. The implications are far-reaching, he said, even for the public. "The immediate effect will be to let loose a veritable intellectual spree of licentious and debauched thinking." For, Bridgman continued, the nonscientist is apt to conclude from the uncertainty principle, not that it stated "the end of meaning," but rather that "there is something beyond the ken of the scientist." In a remarkably prophetic passage, Bridgman wrote:

> This imagined beyond, which the scientist has proved he cannot penetrate, will become the playground of the imagination of every mystic and dreamer. The existence of such a domain will be made the basis of an orgy of rationalizing. It will be made the substance of the soul; the spirits of the dead will populate it; God will lurk in its shadows; the principle of vital processes will have its seat here; and it will be the medium of telepathic communication. One group will find in the failure of the physical law of cause and effect the solution of the age-long prob-

> lem of the freedom of the will; and on the other hand the atheist will find the justification of his contention that chance rules the universe.[1]

Eighty years later, we see that Bridgman was correct: each of these views has indeed been advanced. Bridgman went on to point to a positive side, saying that eventually, we can develop the "new methods of education" to inculcate into people the "habits of thought" required to reshape the thinking we use in "the limited situations of everyday life." The end result, Bridgman concluded, will be salutary:

> [S]ince thought will conform to reality, understanding and conquest of the world about us will proceed at an accelerated pace. I venture to think that there will also eventually be a favorable effect on man's character; the mean man will react with pessimism, but a certain courageous nobility is needed to look a situation like this in the face. And in the end, when man has fully partaken of the fruit of the tree of knowledge, there will be this difference between the first Eden and the last, that man will not become as a god, but will remain forever humble.

Eighty years later, we are still working on acquiring courageous nobility, and remaining humble. But we are also still working on how to talk about the physical interpretation of quantum mechanics, on how it connects with other, more familiar and visualizable features of the world.

Of all the founders of quantum mechanics, Niels Bohr was the most insistent that we should try to fully express the quantum world in the framework of ordinary language and classical concepts. "[I]n the end," as Michael Frayn has Bohr's character say in the play *Copenhagen*, "we have to be able to explain it all to Margrethe," his wife and amanuensis who serves as the onstage stand-in for the ordinary (that is, "classically thinking") person.

Many physicists, finding this task irrelevant or impossible, were satisfied with partial explanations—and Heisenberg argued that the mathematics works: that's enough! Bohr rejected such dodges, and rubbed physicists' noses in what they did not understand or tried to hide. He did not have an answer himself—and knew it—but he had no reason to think one could not be found. His closest answer was the doctrine of complementarity, an ordinary-language way of saying that quantum phenomena behave, apparently inconsistently, as waves or particles depending on how the instruments are set up, and that you need both concepts to fully grasp the phenomena. While this provoked debate among physicists on the "meaning" of quantum mechanics, the doctrine—and discussion—soon all but vanished.

Why? A large part of the answer is that, by 1930, physicists found a perfectly adequate way to represent classical concepts within the quantum framework involving a special abstract mathematical language called Hilbert (infinite-dimensional) space. In this space, the concepts of position and momentum are associated with different sets of coordinate axes that do not line up with each other, resulting in the situation captured in ordinary language terms by *complementarity*.[2] While Bohr used the notion of complementarity to say that quantum phenomena were both particles and waves—somewhat confusingly, and in ordinary language terms—the notion of Hilbert space provided an alternate and much more precise framework in which to say that they are neither. But it was not a language that Margrethe understood; for her, quantum mechanics would have to remain esoteric and she would have to cope with understanding it as best she could. This is what has left the door open for the kinds of fantastic interpretations of meaning for human life mentioned at the beginning of this chapter, and by Bridgman.

What makes the interpretation of quantum mechanics difficult to talk about? It is that we expect a complete theory to fall short of fully describing nature, but in a particular and well-defined way, for it provides a model that is an ideal limit of measurement, with any gaps or discrepancies between it and what we encounter in the laboratory arising from errors and imperfections in the measuring equipment. Many other theories and equations that physicists teach and use have other kinds of gaps and discrepancies, if they omit aspects of nature in the interests of a good approximation. An example is $F = ma$, which leaves out mass-energy conversion in the cause of buying ease of application. These are what we might call "harmlessly fudging or incomplete descriptive theories." Any gaps between the theory and the world are epistemological; that is, they have to do with our knowledge of the world, or the gap between our representations of the world and what it represents.

The uncertainty principle is incomplete in a different sense. It is a mathematical relation, and a feature of the statistical interpretation of the wave function in quantum mechanics. It makes no reference to any underlying physical picture; there are no references to waves or particles, nor to physical experiments. It is not obvious what it refers to, except possibly the clicks of a detector. Yet it is *about* gaps *in* the world itself. These gaps are not epistemological but ontological; having to do not with our knowledge but with the world.

This is strange, but why? It is important to see what the strangeness is *not* due to. The strangeness of the uncertainty principle is not due to the measurement process disturbing the object measured, which would be a feature of any Newtonian theory involving exchange of particles. Nor is it due to the presence of statistics. Rather, the strangeness of quantum mechanics is that quantum formulations are not "about" a real or ideal object in the conventional sense.

In classical physics, deviations of measured quantities from ideal norms are treated independently in a statistically based theory of errors. But the variations—statistical distributions—of quantum measurements are systematically linked in a single formalism. It tells you that all you can know precisely is the width of a distribution, and that you cannot make individual predictions. The superpositions are possibilities in the world, possibilities of observations. The wave formulas of quantum mechanics are thus neither about an ideal object, nor about a real object, but about a special kind of semiabstract object that admits numerous potential experimental realizations in becoming a real object. This special kind of semiabstract object is incomplete if we try to think of it the way we do other more familiar elements of the real world. One needs to add something to the abstract object to bring it into the world, and the choices that one makes regarding what to measure and how to measure it affect what one is measuring. This abstract object can appear wavelike or particle-like, for instance, depending on the kind of situation we put it in, waves and particles being models in visualizable space-time.

This, then, is the "intermediate kind of reality" that Heisenberg said was the price one had to pay to have quantum phenomena. It has the funny kind of incomplete, semiabstract reality that scripts or scores have—they are programs, as it were, for real things in the world (the produced play, the performed music) that require adding a context, and decisions about that context affect the whole of the abstract object. It brings back the role of human purposes and decisions that Newton left out.

The challenge in explaining the meaning of the uncertainty principle to nonscientists lies in trying to explain this new kind of semiabstract object. And it is important to try, for otherwise there will continue to be information loss and distortion in the public understanding of the uncertainty principle.

Heisenberg proved that. Just not mathematically.

CONCLUSION

Bringing the Strange Home

We can bring the strange home, and bring it home with precision.

—Stephen Dunn, *Walking Light: Memoirs and Essays on Poetry*

I have referred to the paths to these equations as journeys, but that metaphor can be misleading. It can mislead because it suggests smooth and steady progress toward a stable and predetermined destination, whereas the path to understanding that culminated in most of these equations was uneven and the travelers often wound up in a different place from where they thought they were heading. The metaphor also falsely suggests that the travelers were spectators taking in a vision of nature, rather than active participants in interactions with it who learned from their changing interactions and often changed their ideas in response.

But the metaphor does capture the way each step of the journey readjusted the perception of the travelers as new sights appeared and others disappeared, and as the overall landscape reorganized around new landmarks. What the travelers thought was important subtly changed as a new world slowly came into view. Such changes were not due to any specific development—to any single distinction, discovery, technique, or person—but to the journey itself. This is what philosophers mean by the historicity of human action. Each group of travelers inherits a landscape, a way of thinking, an accompanying set of dissatisfactions, and a direction to head in to resolve these dissatisfactions, and in the resulting journey the landscape is taken up and transformed. At each step along the way the world seems to have a wild heterogeneity, possessing one order that does not seem inherent in the way the world appears to us—for the order we see in nature is due to our previous explorations and journeys—but to possess hints of another, inherent order that we might be able to see more clearly through inquiry. What Heaviside said of Maxwell's work—"[I]t was only by changing its form of presentation that I was able to see it clearly"—could be said by any of the individuals mentioned in this book. These individuals were discontent with what they saw, had an anticipatory vision of what might be, and the ability to organize an inquiry to seek it (philosophers call this process the hermeneutic circle). There will be no "final" stopping point to the journey, for each new discovery—not to mention changing practical, instrumental, and theoretical contexts—works changes in the landscape. We will never stop being discontent, never stop anticipating, never stop organizing inquiry. Science could not happen in any other way, or it would be trivial or impossible.

Most of the time, however, we care more about the equations, and about the things they help us do, rather than about the journeys that led to them. We tend to pay attention to the part of the world directly beneath our gaze. This is understandable and there are good reasons for doing so. But we can also learn much from studying the journeys—paths from ignorance to knowledge—that the scientific community, and individual scientists, took to these equations.

First of all, we learn that the journeys are very different. Some journeys described in this book are short enough to take an individual just a few minutes. The Pythagorean theorem is an example; it allows someone with no mathematical training not only to grasp it but also to experience the thrill of discovery. Other journeys are extended: the journey to $F = ma$ and $F_g = Gm_1m_2/r^2$ justly can be said to have taken hundreds, even thousands, of years. Some journeys were taken in effect by a community of scientists constantly talking to one another, such as those to $E = mc^2$, to the second law of thermodynamics, and to the uncertainty principle. Other journeys were traveled more or less solo, such as Einstein's to his general equation for gravitation and Schrödinger's to his wave equation, though such individuals in effect carried on conversations with colleagues even when working alone. There is, in short, no single road to discovery.

We also learn that equations are not simply scientific tools but have "social lives," so to speak. We tend to view equations as inert and mute instruments, able to affect the world only when wielded by scientists and engineers. But equations are active and can exert an educational and even cultural force, instructing us about the world and occasionally reshaping the human perception of it. The Pythagorean theorem teaches each new generation of schoolchildren about the meaning of proof, while Newton's law of gravitation taught certain political thinkers about the meaning of laws. The second law of thermodynamics helps keep in check humanity's utopian visions of free energy, while Einstein's $E = mc^2$ and his general equation for gravitation reshaped the human understanding of space and time on a fundamental level. Schrödinger's equation and Heisenberg's uncertainty principle force us to rethink what being a "thing" means.

Yet another thing we learn from these journeys is about the nature of scientific concepts. It is tempting to think that there is some preexisting structure embedded in nature that we are only discovering and translating into mathematical language—that equations are descriptions rather than interpretations or creations. But how we translate depends on the journey we have already taken, on our dissatisfactions with it, and on how we responded to those dissatisfactions. We "fall up," to adapt a phrase of George Steiner. It is thus misleading to picture science as proceeding solely by scientists producing new concepts, then testing and revising them. Two things are wrong with this picture. One is that the meaning of one concept depends on the meaning of all the others; a concept is one element of that fishbowl-like world that Newton discovered at the heart of the world we live in, and needs everything else in that fishbowl for its meaning. Testing one concept thinking you are testing it and not everything else is like asking, Is New York to the right or left of Boston? without knowing where you are; without having the rest of the map. And we not only need the rest of the fishbowl, but the rest of our experience of the world as well. A scientific concept that we trust is really a concept plus that experience, and when our experience changes—new practices, new technologies—so does how the concept applies to the world. That's why concepts never stay put, and always change or are being elaborated; a concept that tests right at one time can be inadequate at another. There is no right way to say something that does not include our experience with it. Concepts are thus not determinative but indicative; they "point" to something based on our experience, in full awareness that what they point to is going to change with further inquiry. Philosophers call this "formal indication." Concepts are formal because we can evaluate them rigorously and test them as being adequate or inadequate; they belong to a closed system. Concepts are indicative because they point to and depend on other things for their adequacy—all our experiments and definitions and technology and open-ended connection to the world—and when these change so can the formal elements as well. In fact, we expect that it will. Historian of science Peter Galison has a wonderful description of this. It is the theorist's experience, he writes, that:

> You try adding a minus sign to a term—but cannot because the theory then violates parity; you try adding a term with more particles in it—forbidden because the theory now is nonrenormalizable and so demands an infinite number of parameters; you try leaving a particle out of the theory—now the law has uninterpretable probabilities; you subtract a different term—all your particles vanish into the vacuum; you split a term in two—now charge is not conserved; and you still have to satisfy conservation laws of angular momentum, linear momentum, energy, lepton number, and baryon number. Such constraints do not all issue axiomatically from a single,

governing theory. Rather, they are the sum total of a myriad of interpenetrating commitments of practice. Some, such as the conservation of energy, are over a century old. Others, such as the conservation of parity, survived for a very long time before being discarded. And yet others, such as the demand for naturalness—that all free parameters arise in ratios on the order of unity—have their origin in more recent memory. Some are taken by the research community to present nearly insuperable barriers to violation, while others merely flash a yellow cautionary light on being pushed aside. But taken together, the superposition of such constraints makes some phenomena virtually impossible to posit, and others (such as black holes) almost impossible to avoid.[1]

Yet another thing we learn from these journeys is that science is a deeply affective process. Those who do not realize this, or who think that scientific experience involves a dry conceptual part plus a separate emotional part, do not understand science or human creativity. It is possible—and indeed useful for some purposes—to divide up the scientific process into a conceptual part and an affective part, but this is an artificial model, something that comes afterward. Studying these journeys allows us to bore underneath the levels of abstraction that conceal how science truly works. We encountered the role of dissatisfaction, for instance, in many of these journeys, and also saw episodes of curiosity, consternation, bafflement, and wonder. We saw the difference between expectation and alertness—between scientists who expected something and could only take notice when that expectation was fulfilled, and scientists who were alert in the sense that they were prepared to hear something more than what they expected. We encountered affects not only in what motivates discovery, but also in the scientists' response to it. The affective response to a discovery is not simply "OK, I get it now, this belongs here and that over there, I had it wrong and I get it now," but something much more nuanced and powerful. Nor is the affective response limited to discovery. As Leon Lederman wrote, to pin one's hopes on making a discovery that will bring fame and fortune "is not a life." He continued, "The fun and excitement must be daily—in the challenge of creating an instrument and seeing it work, the joy of communicating to colleagues and students, the pleasure of learning something new in lectures, corridors, and journals."[2]

But our wonder is not only at what we have learned, but at something still more profound. In certain moments of wonder, we glimpse the connection between ourselves and nature; we glimpse the mutability of nature and our role in it. We experience that nature could be otherwise—more, that it *was* otherwise until a moment ago, and for all we know it could change in the future. In such moments, we experience an Emersonian moment of a higher thought in the middle of the existing one, a more profound feeling befalling us that we experience at once as new and old, surprising and familiar, there and not there before us, uncanny and domestic.

Shortly before finishing this book, I found myself struggling to describe the project to an eminent, elderly physicist who expressed little sympathy for books about science accessible to nonscientists. No magic for him! The equations, when fully grasped, seem so obvious, or so complete, or so logical that, once grasped, we cannot imagine not having known them. He approached science the way he thought a purely professional workman should, and urged me to do likewise. "Such equations," he told me, "would not be wonderful if people realized how trivial they are. You should help them do so."

I could have hugged the old man. He helped me put my finger on what I was trying to do. Which was just the opposite—to show how equations are *not* trivial, to recover the dissatisfactions that led us to seek them, and to restore the wonder to the moment when we first grasped them. The wonder at the moment when they arrived, seemingly simultaneously discovered and invented, when they seemed more concise statements of what we already know, something (like the Pythagorian theorem) so secure that it seems that it was already "in us" and simply remembered. Scientists of the sort as my elderly physicist acquaintance tend to be focused on the formal (what he meant by "trivial") part, whereas philosophers and other scholars in the humanities tend to focus on the other part. It ought to be possible to have both parts at once: the sciences and the humanities together, anosognosia cured, Twain's young and old pilots viewing the water, the slave boy with his eye on the diagram and Socrates with his on human life, the formal part and the meaningful, affective part put back together in the originary unity from which they sprang. If we can, we will recover the wonder of Richard Harrison's child discovering $1 + 1 = 2$, and view equations as the key "not only to what was wonderful in the outside world, but what was wonderful in him and all of us." Such a moment would be a fully human response to the world.

NOTES

INTRODUCTION

1 Modern astrologers seem strangely untroubled by, and even ignorant of, the fact that constellations do not come in neat packages—and that the sun and planets pass by them, and sometimes by entirely different constellations, at different times than confidently asserted by the dates given in newspaper horoscopes. Someone ought to file a malpractice claim.

2 See I. Bernard Cohen, *The Triumph of Numbers: How Counting Shaped Modern Life* (New York: W. W. Norton, 2005).

3 In response to this need to use something to stand for numbers or other things—symbols—the ancient Egyptians, Babylonians, and Greeks developed different ways of symbolizing numbers and quantities. Much ancient mathematics then consisted of solving specific cases and inviting the reader to generalize. For example, the famous Rhind papyrus, an Egyptian manuscript from about 1650 BC, contains what amount to rudimentary equations, based on examples, for figuring out the areas of triangles, rectangles, circles, and the volumes of prisms and cylinders. The papyrus also demonstrated solutions for practical problems, such as how to determine equalities between loaves of bread of different consistencies and different amounts of barley. It even discussed exemplary problems that are not practical but conceptually interesting, such as the following: "There are seven houses; in each house there are seven cats; each cat kills seven mice; each mouse has eaten seven grains of barley; each grain would have produced seven 'hekat.' What is the sum of all the enumerated things?" Many different versions of this problem have cropped up ever since, such as

the Mother Goose rhyme "The Man from St. Ives"—who had seven wives, each of whom had seven sacks, each of which contained seven cats, each of which had seven kittens. In their equations, the Egyptians used symbols that consisted of hieroglyphs looking like pairs of legs that seem to be walking in the direction the book is written for addition, or in the opposite direction for subtraction.

4 A. N. Whitehead and B. Russell, *Principia Mathematica*, vol. 1 (Cambridge: Cambridge University Press, 1957), p. 362: "From this proposition it will follow, when arithmetical addition has been defined, that $1 + 1 = 2$."

5 Today, equations are classified in several different ways. One is by their degree, or the character of its biggest exponent. In linear equations, so-called because they describe lines (examples include $4x + 3y = 11$ and $y = 2x + 1$), the unknown numbers x or y are not raised to any power and are said to be of the first degree. When the unknown is squared, the equation is called quadratic; when cubed, it is a cubic equation; after that it is an equation of the fourth, fifth, sixth degrees, and so on. And when the solution to an equation is not a number but a function—when it is said to contain "derivatives"—it is called a differential equation.

6 In Isaac Newton, *The Principia: Mathematical Principles of Natural Philosophy*, trans. I. B. Cohen and Anne Whitman (Berkeley: University of California Press, 1999), p. 391.

7 This has led some people to compare equations and poems. Both involve special uses of language often above the heads of untrained readers that seek to express truths concisely and with precision, and that allow us to understand otherwise inaccessible things, changing our experience in the process. Equations "state truths with a unique precision, convey volumes of information in rather brief terms, and often are difficult for the uninitiated to comprehend," writes Michael Guillen in his book *Five Equations that Changed the World: The Power and Poetry of Mathematics*. Guillen adds, "And just as conventional poetry helps us to see deep within ourselves, mathematical poetry helps us to see far beyond ourselves." Graham Farmelo, the editor of *It Must Be Beautiful: Great Equations of Modern Science*, likewise compared equations and poems. Both, he noted, are composed of abstractions with which we address the world, even though many individual terms do not have specific referents. While "poetry is the most concise and highly charged form of language," Farmelo writes, equations are "the most succinct form of understanding of the aspect of physical reality they describe."

Many other differences exist, of course, between poems and equations. Equations can seem fearful. For they are not only beyond our understanding, but refer to powers beyond our control—which can make us feel helpless and resentful. Poems generally refer directly to the intuitive human experience of the surrounding world, invoke more than inform, and do so in a way that can have an impact on that intuitive experience. Equations, by contrast, do not refer to direct human experience, but to specially defined quantities—such as acceleration, energy, force, mass, the speed of light, to name a few—that are measured in laboratories. They cannot be plucked or dug up from anywhere, ordered from a catalogue, or held in your hand like apples and balls. And equations have a special structure that poems lack—they state that one group of these quantities is equal to (or greater to or less than, in their looser sense) another.

Such quantities—what equations refer to—are not always easy to identify. Consider the old saw about the Army captain seeking to hire a lieutenant, who posed to each of the three candidates the same question: "How much is 1 + 1?"

Candidate 1 answered, "Two, of course."

Candidate 2 answered, "Well, that all depends on what 1 represents. It might be a vector, in which case its value could be anything from 0 to +2."

And Candidate 3 answered, "How much would you like it to be?"

The predictable punch line, of course, is that the job goes to Candidate 3. Part of the joke—Candidate 2's contribution—relies on an equivocation: between a number and the magnitude of a vector. But the buildup shows that the ties between equations and the world are not as simple as it appears. Still, specifying the relations among specially defined quantities allows equations to transform our encounters with the world in several ways—by pointing out new things, by giving us more power, and by reorganizing the way we see. Poems don't do it that way.

8 Frank Wilczek, "Whence the Force of F = ma? I: Culture Shock," *Physics Today*, October 2004, pp. 11–12; "Whence the Force of F = ma? II: Rationalizations," *Physics Today*, December 2004, pp. 10–11; "Whence the Force of F = ma? III: Cultural Diversity," *Physics Today*, July 2005, pp. 10–11.

CHAPTER ONE "The Basis of Civilization"

1 John Aubrey, *Brief Lives*, ed. Richard Barber (Great Britain: Boydell Press, 1982), p. 152.

2 This story might seem too apocryphal to be true, a retrospective "Eureka!" moment, but most biographers believe it. Often we realize only later the significance of a moment whose meaning we are only dimly aware of at the time. Hobbes's recent biographer A. P. Martinich (*Hobbes: A Biography*, pp. 84–85) argues forcefully for the truth of the story. Martinich adds, "The importance of geometry on Hobbes's philosophy can hardly be exaggerated. . . . What came to impress Hobbes was not so much the axioms, theorems, and proofs of geometry itself, but the method of connecting one thing with another on a foundation that could not be doubted. It was the method, not the substance, of geometry that staggered him."

3 He never became a true professional, though, and fell into traps of the sort that enthusiastic amateurs often do. These included pursuing impossible problems like trying to square the circle, trisect an angle, and double a cube, each of which Hobbes erroneously thought he had achieved.

4 Leo Strauss, *The Political Philosophy of Hobbes: Its Basis and Its Genesis* (Chicago: University of Chicago Press, 1959), p. 29.

5 Reid McInvale, "Circumambulation and Euclid's 47th Proposition," http://www.io.com/~janebm/summa.html (accessed April 11, 2008). See also James Anderson, *The Constitutions of the Free-Masons* (1723): "[T]he Greater Pythagoras, prov'd the Author of the 47th Proposition of Euclid's first Book, which, if duly observ'd, is the Foundation of all Masonry, sacred, civil, and military. . . ." *Little Masonic Library* [rev. ed.], vol. 1 (Richmond, VA: Macoy, 1977), pp. 203–4.

6 O. Neugebauer and A. Sachs, "Mathematical Cuneiform Texts," in *American Oriental Series*, vol. 29 (New Haven: American Oriental Society, 1945), p. 38; Eleanor Robson, "Neither Sherlock Holmes nor Babylon: A Reassessment of Plimpton 322," *Historia Mathematica* 28 (2001), pp. 167–206.

7 The Baudhāyana, for instance, says that "the diagonal of an oblong produces by itself both the areas which the two sides of the oblong produce separately" [quoted in David Smith, *History of Mathematics*, vol. 1 (New York: Dover, 1958), p. 98], but simply declares this as a fact without further justification. "[W]e must remember," writes one scholar, "that they were interested in geometrical truths only as far as they were of practical use, and that they accordingly gave to them the

most practical expression" [G. Thibaut, *The Śulvasūtras* (Calcutta: Papatist Mission Press, 1875), p. 232].

8 Christopher Cullen, *Astronomy and Mathematics in Ancient China: The Zhou Bi Suan Jing* (Cambridge: Cambridge University Press, 1996), p. xi. But as Cullen observes, "the process is more verbal than computational," and "to illustrate it by a carefully labelled Euclidean diagram when none is referred to in the text is perhaps only a way of misleading oneself" about what the author is up to, for "nothing worthy of being called computation is involved" (p. 80). The height of the sun, by the way, is 80,000 li, or about 40,000 kilometers or 24,000 miles. A later Chinese text called the *Ziu Zhang Suan Shu* (Nine Chapters on the Mathematical Art, from about AD 250), has the rule somewhat more explicitly treated. The *Ziu Zhang*'s concerns are mainly practical; its first chapter is "field measurement," while later chapters are on canals, taxation, and other matters. Its ninth and final chapter is "Kou ku," or "base and altitude," *kou* or "leg" meaning the short side of a right triangle, *ku* or "thigh" the long side (*hsien* meant the hypotenuse or line strung between two points). The chapter contained twenty-four problems on properties of right triangles. But "proof is not their preoccupation," says historian G.E.R. Lloyd. "[T]heir style of mathematical reasoning has more to do with exploring analogies and common structures (in groups of problems, procedures, formulae) than with demonstration as such—a style that itself remains close to that favoured in other genres, including poetry, also remarkable for its interest in correlations, complementarities, parallelisms. The contrast, here, with the Greek opposition of proof and persuasion—fueled by the quest for incontrovertibility—could hardly be more striking" [G.E.R. Lloyd, *Demystifying Mentalities* (Cambridge: Cambridge University Press, 1990), pp. 121–22]. "And that is the main point," Cullen observes after noting Lloyd's comments: "even when an ancient Chinese mathematician gives a proof, it is not very important to him in comparison with his real aim of explaining the use of the methods he is expounding to solve specific problems." He adds (p. 89), "Why should it be otherwise?"

9 The hypotenuse rule was well known to Greek authors who lived a century or so after Pythagoras, and none of these authors attributes it to Pythagoras. Aristotle—who is good about attributing credit where it's due—also knew the proof, but says nothing of any tie to Pythagoras. The idea of a proof began to emerge in the fifth century BC, and culminated in the fourth with Plato's discussion of the distinction between persuasion and demonstration, with Aristotle's discus-

sion of the nature of proof, and finally with Euclid's *Elements*, a book that presents mathematical knowledge entirely in the form of proofs. There remains a major difference between the early writings exhibiting knowledge of mathematical rules in obtaining practical results, and the later Greek idea of formal proof. "Practice is one thing, having the explicit concept another," writes Lloyd. "[T]o give a formal proof of a theorem or proposition requires at the very least that the procedure used be exact and of general validity, establishing by way of a general, deductive justification the truth of the theorem or proposition concerned" (Lloyd, *Demystifying Mentalities*, pp. 73, 74). This, Lloyd continues, was first defined, as far as we know, not just in Greece but anywhere, by Aristotle. Though some individuals make claims for earlier discoveries of the Pythagorean theorem, in Mesopotamia, India, and China, for instance, "[I]n the key texts we find no distinction observed between *exact* procedures and approximate ones. Both are used apparently indiscriminately, and that suggests that their authors were not concerned with *proving* their results at all, but merely with the concrete problems of altar construction" (p. 75). It is true that the first proof of the hypotenuse formula is traditionally ascribed to Pythagoras (ca. 569–475 BC), by authors who lived about half a millennium later, around the time of the birth of Christ. But this attribution may well be, as Lloyd remarks, the result of the tendency of "the late Greek commentators to make overoptimistic attributions of sophisticated ideas to the heroic founders of Greek philosophy" (p. 80). The culprit seems to be a certain Apollodorus, about whom nothing is known except his remark that Pythagoras sacrificed oxen upon discovering a "famous theorem." Apollodorus's remark was then relied on by many other authors—who include Plutarch, Athenaeus, Diogenes Laertius, Porphyry, and Vitruvius. Some authors embellished the story, while others express skepticism about the sacrifice, given that the Pythagoreans had strictures against rituals in which blood was shed. "What is both uncontroversial and of first rate importance for the subsequent development of Greek science," Lloyd concludes (p. 87), "is the role that Euclid's *Elements* itself had as providing the model for the systematic demonstration of a body of knowledge. Thereafter proof *more geometrico* became all the rage, and not just in geometry, but also for example in optics, in parts of music theory, in statics and hydrostatics, in parts of theoretical astronomy, and not just in the would-be exact sciences, but in some of the life sciences as well."

10 To name one, Otto Neugebauer, who first deciphered the Pythago-

rean triplets of Plimpton 322, cited old Babylonian tablets as "sufficient proof that the 'Pythagorean' theorem was known more than a thousand years before Pythagoras." Otto Neugebauer, *The Exact Sciences in Antiquity* (Providence: Brown University Press, 1993), p. 36.

11 Francis M. Cornford, *Before and After Socrates* (Cambridge: Cambridge University Press, 1972), pp. 72–73.

12 *American Mathematical Monthly* 1, no. 1 (January 1894), p. 1.

13 Elisha S. Loomis, *The Pythagorean Proposition: Its Proofs Analyzed and Classified*. Publ. by The Masters and Wardens Association of the 22nd Masonic District of the Most Worshipful Grand Lodge of Free and Accepted Masons of Ohio, 1927; and *The Pythagorean Proposition: Its Demonstrations Analyzed and Classified* (Ann Arbor, MI: Edwards Brothers, 1940). He ended the first book thus: "FINAL THOUGHT: Is it an all-embracing truth? The generalization of the Pythagorean Theorem so as to conform to and include the data of geometries other than that of Euclid, as was done by Riemann in 1854, and later, 1915, by Einstein in formulating and positing the general theory of relativity, seems to show that the truth implied in this theorem is destined to become the fundamental factor in harmonizing past, present and future theories relative to the underlying laws of our universe."

14 For instance, geometric proof 32 of the second edition "is credited to Miss E. A. Coolidge, a blind girl"; geometric proof 68 is "the first ever devised in which all auxiliary lines and all triangles used originate at the middle point of the hypotenuse of the given triangle. It was devised and proved by Miss Ann Condit, a girl, aged 16 years, of Central Junior-Senior High School, South Bend, Ind., Oct. 1938. This 16-year-old girl has done what no great mathematician, Indian, Greek, or modern, is ever reported to have done"; geometric proof 69 "is original . . . devised by Joseph Zelson, a junior in West Phila., Pa., High School, and sent to me by his uncle. . . . It shows a high intellect and a fine mentality," and Loomis adds that "this proof and the one before "are evidences that deductive reasoning is not beyond our youth"; geometric proofs 252–55 "show high intellectual ability, and prove what boys and girls can do when permitted to think independently and logically"; and regarding algebraic proof 93, Loomis remarks that it is "by Stanley Jashemski, age 19, of Youngstown, O., June 4, 1934, a young man of superior intellect."

15 Loomis, 1927 edition, p. 99.

16 Loomis, 1940 edition, p. 269.

17 Eli Maor, *The Pythagorean Theorem: A 4,000-Year History* (Princeton: Princeton University Press, 2007), p. xiv.

18 Galileo, *Galileo on the World Systems*, trans. M. A. Finocchiaro (Berkeley: University of California Press, 1997), p. 97.

19 G.W.F. Hegel, *Hegel's Philosophy of Nature*, vol. 1, ed. and trans. M. J. Petry (New York: Humanities Press, 1970), p. 228.

20 I thank my colleague David Dilworth for this observation.

INTERLUDE Rules, Proofs, and the Magic of Mathematics

1 Oliver Byrne, ed. *The First Six Books of the Elements of Euclid* (London: William Pickering, 1847). It is available online at http://www.math.ubc.ca/people/faculty/cass/Euclid/byrne.html.

2 David Socher, "A Cardboard Pythagorean Teaching Aid," *Teaching Philosophy* 28, 2005, pp. 155–61.

3 George MacDonald Fraser, *Quartered Safe Out Here: A Recollection of the War in Burma* (London: HarperCollins, 1992), p. 150.

4 From pp. 9–11 in the opening autobiographical sketch of *Albert Einstein: Philosopher-Scientist*, ed. Paul Arthur Schilpp (London: Cambridge University Press, 1970).

CHAPTER TWO "The Soul of Classical Mechanics"

1 Thus the subjective experience individuals have of themselves as a bodily center of action was metaphorically projected into nonhumans as one of their properties. This illustrates what philosopher Maxine Sheets-Johnstone calls "the living body serv[ing] as a semantic template," a process which, she points out, is key to the emergence of many early scientific concepts. It is a classic case of the use of analogical thinking, or the use of the familiar to understand the unfamiliar. What is remarkable in this instance is that the familiar has its basis in the tactile-kinesthetic experiences of bodily life; thus, in a corporeal template. For the connection between early ideas of force and religious ideas, see Max Jammer, *Concepts of Force: A Study in the Foundations of Dynamics* (Cambridge: Harvard University Press, 1957), chapter 2.

2 Nearly all modern editions of Aristotle include the pagination of a standard edition published in 1831, and when making references it is standard practice—and highly efficient given the numerous editions and translations—to cite the name of Aristotle's book followed by the chapter and, sometimes, line. This reference is to *On the Heavens*, Book I, ch. 3, lines 270 b 13–17.

3 That unchanging things in the heavens, he concluded, move "with a ceaseless, circular movement" is clear "not only to reason, but also in fact." *Metaphysics*, 1072a21.

4 The full set of such rules of thumb can be found in *Physics*, Book VII, ch. 5, *Physics*, Book VIII, ch. 10, and *On the Heavens*, Book I, ch. 7. They include: "Half the force will move the same body half the distance in the same time"; "The same force will move a body half as heavy twice the distance"; "Twice the resistance halves the distance"; "The thicker the medium, the more slowly a body falls in it"; "The heavier the body, the faster it falls." It is tempting nowadays to express these rules mathematically. Later commentators, looking back at Aristotle from the vantage of thousands of years later, paraphrased Aristotle's statements on motion as follows: Given the same time and force, the distance traversed by an object is inversely proportional to the resistance; and given the same distance and force, the time is directly proportional to the resistance. Often these were simplified still further in mathematical notation, combining distance and time as velocity (V), and representing F as force and R as resistance:

$$V \propto F/R$$

(velocity is proportional to force divided by the resistance)

and

$$V \propto W/R$$

(velocity is proportional to weight divided by the resistance)

But that would be misleading, and misrepresent what he saw. He knew there were exceptions and even areas where the rules did not apply. He knew, for instance, that the connection between force and speed was not smoothly varying—for while 50 people could push a ship half as far in the same time as 100 people, 1 person couldn't push it at all. If the force equaled the resistance, the movement is clearly zero, but his rules suggested otherwise. And he believed that the speed of an object increases as it gets closer to its natural place, which is not reflected in his rules.

5 Aristotle had no sense of uniform motion. He was less interested in the stages of motion—uniform, accelerated, uniformly accelerated—than in where the moved object came from and where it was heading. Thus he had no idea of speed as a particular instant versus average speed. Speed, to him, was overall speed, the time it takes something to complete a movement, and he noted that some movements take

more time than others. "Velocity as a technical scientific term to which numerical values might be assigned," Lindberg notes, "was a contribution of the Middle Ages." David C. Lindberg, *The Beginnings of Western Science: The European Scientific Tradition in Philosophical, Religious, and Institutional Context, 600 B.C. to A.D. 1450* (Chicago: University of Chicago Press, 1992), p. 60.

6 *On the Heavens*, Book I, ch. 8.

7 Ibid., Book III, ch. 2.

8 Aristotle's two views here are from the fourth and eighth books of the *Physics*. For an analysis see Marshall Clagett, *The Science of Mechanics in the Middle Ages* (Madison: University of Wisconsin Press, 1959), pp. 505–9.

9 See Clagett, *The Science of Mechanics*, pp. 258–61.

10 "Impetus is a thing of permanent nature distinct from the local motion in which the projectile is moved. . . ." Quoted in Lindberg, *The Beginnings of Western Science*, p. 303.

11 "[I]t is unnecessary to posit intelligences as the movers of celestial bodies. . . . For it could be said that when God created the celestial spheres, He began to move each of them as He wished, and they are still moved by the impetus which He gave to them because, there being no resistance, the impetus is neither corrupted nor diminished." Quoted in Simon Oliver, *Philosophy, God and Motion* (New York: Routledge, 2005), p. 152.

12 He proved, for instance, the mean speed theorem, that a uniformly accelerating body (a car accelerating from 0 to 60 mph in a minute, we might say) covers the same ground as a body moving at the mean uniform speed (30 mph for a minute). From Jammer, *Concepts of Force*, p. 66.

13 "[I]t is necessary that points, lines, and surfaces, or their properties be imagined. . . . Although indivisible points, or lines are non-existent, still it is necessary to feign them." Marshall Clagett, ed. and trans., *Nicole Oresme and the Medieval Geometry of Qualities and Motions* (Madison: University of Wisconsin Press, 1968), p. 165.

14 See, for instance, I. B. Cohen, *The Triumph of Numbers: How Counting Shaped Modern Life* (New York: W. W. Norton, 2005).

15 For Galileo's use of force see Richard Westfall, *Force in Newton's Physics: The Science of Dynamics in the Seventeenth Century* (New York: Elsevier, 1971), chapter 1 and Appendix A.

16 Westfall, *Force in Newton's Physics*, pp. 41–4`2.

17 Jammer, *Concepts of Force*, p. 120.

18 I. Bernard Cohen, "Newton's Second Law and the Concept of

Force in the *Principia*," in *The Annus Mirabilis of Sir Isaac Newton 1666–1966* ed. Robert Palter (Cambridge, MA: MIT Press, 1971), p. 171.

19 Isaac Newton, *The Principia: Mathematical Principles of Natural Philosophy*, trans. I. Bernard Cohen and Anne Whitman (Berkeley: University of California Press, 1999), p. 407.

20 Newton, *Principia*, p. 409.

21 Herbert Butterfield, quoted by Oliver, *Philosophy, God, and Motion*, p. 168.

22 Cohen, "Newton's Second Law," p. 143. Cohen goes on to describe the evolution of Newton's ideas leading up to the *Principia*, saying that "[T]he Second Law may serve as a particularly fascinating index to Newton's achievement in the *Principia* because it reveals to us how Newton was able to generalize his physics from the phenomenologically based dynamics of collisions and blows to the debatable realm of central forces, of gravitational attraction, and hence of continuous forces generally" (p. 160).

23 As Newton writes in the Preface, the purpose of the book is "to discover the forces of nature from the phenomena of motions and then to demonstrate the other phenomena from these forces." He will use the forces, for instance, to deduce "the motions of the planets, the comets, the moon, and the sea." *Principia*, p. 382.

24 Voltaire to Pierre-Louis Moreau de Maupertuis, October 1932, in *Voltaire's Correspondence*, vol. 2, ed. T. Besterman (Geneva: Voltaire Institute and Museum, 1953), p. 382.

INTERLUDE The Book of Nature

1 Galileo, "The Assayer," in *Discoveries and Opinions of Galileo*, trans. S. Drake (New York: Doubleday, 1957), pp. 237–38.

CHAPTER THREE "The High Point of the Scientific Revolution"

1 Another interesting idea came from the mathematician Pappus of Alexandria (third century AD), who proposed a way of treating gravity as if it were an Aristotelian "pulling": find out how much pull it takes to move a weight on a plane, he said, and then tilt the plane to find out how much more pull it takes to move the weight upward. Jammer, *Concepts of Force*, p. 41.

2 Lindberg, *The Beginnings of Western Science*, p. 275.

3 See Lynn Thorndike, "The True Place of Astrology in the History of Science," *ISIS* 46 (1955), p. 273.

4 Nicoletto Vernias, *De gravibus et levibus*, Venice 1504, cited in Jammer, *Concepts of Force*, p. 67.

5 "I think that gravity," Copernicus wrote, "is nothing else than a certain natural appetition given to the parts of the earth by divine providence of the Architect of the universe in order that they may be restored to their unity and to their integrity by reuniting in the shape of a sphere. It is credible that the same affection is in the sun, the moon, and other errant bodies in order that, through the agency of this affection, they may persist in the rotundity with which they appear to us." *De Revolutionibus*, book 1, chapter 9.

6 The idea that the planets travel elliptical paths was a fundamental break with long-held principles. Trust—in Tycho and in Copernicus's heliocentric idea—enabled Kepler to appreciate the significance of the excess of Brahe's data over the theory, and to question assumptions held since ancient times. Trust made the discrepancy meaningful, and directed his suspicion to the right place. This was not the first time that trust played a central role in a major scientific discovery, nor will it be the last.

7 E. A. Burtt, *The Metaphysical Foundations of Modern Science* (Garden City, NY: Doubleday, 1954), p. 64.

8 Johannes Kepler, *Mysterium Cosmographicum*, trans. E. J. Aiton (Norwalk, CT: Abaris Books, 1999), p. 203.

9 In 1666, Italian physiologist Alfonso Borelli (1608–1679) proposed an explanation for the motions of the moons of Jupiter, involving an interaction of several forces, that invited application to planetary motions, suggesting that the same laws govern both the planets and their moons, and the sun and the solar system.

10 Carl B. Boyer, "Boulliau, Ismael," entry in the *Dictionary of Scientific Biography*, vol. 2 (New York: Scribner's, 1970), pp. 348–49.

11 Robert Hooke, "Lectiones Cutlerianae," in R. T. Gunther, *Early Science in Oxford*, vol. 8, 1908, pp. 27–28.

12 Hooke to Newton, November 24, 1679. Isaac Newton, *The Correspondence of Isaac Newton, Vol. II, 1676–1687*, ed. H. W. Turnbull (Cambridge: Cambridge University Press, 1960), p. 297.

13 Writes Newton's biographer Westfall, "Few periods have held greater consequences for the history of Western science than the three to six months in the autumn and winter of 1684–5." R. S. Westfall, *Never at Rest: A Biography of Isaac Newton* (New York: Cambridge University Press, 1988), p. 420.

14 Feynman, *Lectures on Physics*, tape 13, no. 1, side 1.
15 I. Bernard Cohen, *Scientific American*, March 1981.
16 "Newton was the one who elevated Kepler's law of areas to the status it enjoys today." Cohen, *Scientific American*, March 1981, p. 169.
17 Quoted in I. Bernard Cohen, *Birth of a New Physics* (New York: W. W. Norton, 1985), p. 151.
18 Ibid., p. 236.
19 This was a remarkable development, but one whose pattern recurs in the history of science: Newton's early work had been motivated by Kepler's laws; he assumed that they were an accurate description of nature and they led him to a deep insight, yet the insight implied that Kepler's laws were wrong, and allowed Newton to predict deviations from Kepler's laws. This development illustrates how the human mind bootstraps itself in science, engaging in a back-and-forth interaction between two realms—our experience of nature and our models of it, how nature appears and the concepts through which we encounter it, and the way that this process changes both how nature appears and our concepts. Philosophers call such a process *hermeneutics*, a fancy term for interpretation, but it merely expresses a basic scientific procedure, a process often hidden because we tend to fix our eyes on nature rather than on the process. But without it, science would be trivial or impossible.
20 As he writes to Bentley, "Gravity must be caused by an agent acting constantly according to certain laws; but whether this agent be material or immaterial, I have left to the consideration of my readers." I. Newton *The Correspondence of Isaac Newton, Vol. III, 1688–1684*, ed. H. W. Turnbull (Cambridge: Cambridge University Press, 1961), p. 254.
21 This made historian Marjorie Nicolson wonder whether "Newton felt that his formulation of the law of gravitation was not so much the beginning of something new as the climax of something very old." She continued: "Here was the ultimate proof that the microcosm does reflect the macrocosm, that there is a repetition, interrelationship, interlocking between parts and whole, long surmised by classical, medieval, Renaissance scientists, poets, mystics: the law that governs the planets and restrains the stars in their macrocosmic courses is the same law that controls the falling of a weight from the Tower of Pisa or the feather from the wing of a bird in the little world, of which man still remains the center." Marjorie Hope Nicolson, *The Breaking of the Circle: Studies in the Effect of the "New Science" Upon Seventeenth-Century Poetry* (New York: Columbia University Press, 1960), p. 155.
22 Westminster, 1728. The poem is discussed in I. Bernard Cohen,

Science and the Founding Fathers: Science in the Political Thought of Thomas Jefferson, Benjamin Franklin, John Adams, and James Madison (1995), pp. 285–87.

23 H. Saint-Simon, in *Henri Saint-Simon: Selected Writings*, ed. K. Taylor (London: Croom Helm, 1975), pp. 78–79.

INTERLUDE That Apple

1 See D. McKie and G. R. de Beer, "Newton's Apple," *Notes and Records of the Royal Society of London* 9 (1951), pp. 46–54.

2 Westfall, *Never at Rest: A Biography of Isaac Newton* (New York: Cambridge University Press, 1988), p. 155.

3 William Stukeley, *Memoirs of Sir Isaac Newton's Life*, ed. A. Hastings White (London: Taylor and Francis, 1936), pp. 19–20.

4 E. N. da C. Andrade, *Sir Isaac Newton, His Life and Work* (New York: Doubleday Anchor, 1950), p. 35.

5 I. Bernard Cohen, "Newton's Discovery of Gravity," *Scientific American*, March 1981, p. 167.

CHAPTER FOUR "The Gold Standard for Mathematical Beauty"

1 Ed Leibowitz, "The Accidental Ecoterrorist," *Los Angeles* magazine, May 2005, pp. 100–105, 198–201.

2 Quoted in Carl A. Boyer, *A History of Mathematics* (Princeton: Princeton University Press, 1985), p. 482.

3 Marquis de Condorcet, "Eloge to Mr. Euler," trans. J. Glaus, www.groups.des.st-and.ac.uk/~history/Extras/Euler_elogium.html.

4 Martin Gardner, *The Unexpected Hanging and Other Mathematical Diversions* (Chicago, University of Chicago Press, 1961) has an excellent chapter (3) on *e*.

5 R. Feynman, R. Leighton, and M. Sands, *The Feynman Lectures on Physics*, vol. 1 (New York: Addison-Wesley, 1963) has excellent sections (22-5 and 22-6) on imaginary numbers and imaginary exponents.

6 Quoted in Boyer, *History of Mathematics*, p. 493.

7 Condorcet, "Eloge to Mr. Euler."

8 David M. Burton, *The History of Mathematics* (New York: McGraw-Hill, 1985), p. 503.

9 We can retrieve 2^x for any arbitrary x by multiplying x by the natural logarithm $\ln(2)$, and then exponenting: $2^x = e^{x\ln(2)}$.

10 Leonhard Euler, *Introduction to Analysis of the Infinite*, book 1, trans. J. D. Blanton (New York: Springer, 1988), p. 112. Euler first published this in *Miscellanea Berolinensia* 7 (1743), p. 179.

11 G. H. Hardy, P. V. Seshu Aiyar, and B. M. Wilson, eds., *Collected Papers of Srinivasa Ramanujan* (New York: Chelsea Publishing Company, 1962), p. xi.

12 This wonderful way of representing Euler's formula is presented in L.W.H. Hull's note, "Convergence on the Argand Diagram," *Mathematical Gazette* 43 (1959), pp. 205–7. Many thanks to George W. Hart for pointing this out, and for suggesting the different fonts.

13 Herbert Turnbull, quoting Felix Klein, "The Great Mathematicians," in *The World of Mathematics*, vol. 1, ed. James R. Neuman (New York: Simon and Schuster, 1956), p. 151.

14 Yet this is not the most general expression. Mathematicians have sometimes argued, for instance, whether π is defined most economically. That is, given all the 2πs found in math and science, and the vast simplification that results by making π radians the length around a unit circle, are there not beauties and economies to making the fundamental constant here the ratio of the circumference to the radius? To put it another way, are there any examples of places where the beauties and economies lie with π? The most obvious candidate is $e^{i\pi} + 1 = 0$. At first sight, it would seem to subtract from the elegance of this equation to become $e^{i\pi/2} + 1 = 0$. Yet mathematicians have discovered a twist. Suppose we use the symbol ψ to designate 2π. Then we can write a more beautiful and economical formula, of which Euler's formula is just a special case: $e^{i\psi/n} = \sqrt[n]{1}$. This is more general, because one of the square roots of 1 is -1. Euler's formula is a special case of this equation similar to the way that the Pythagorean theorem is a special case of the law of cosines.

INTERLUDE Equations as Icons

1 Larry Wilmore, quoted in *The New York Times*, April 15, 2007, section 4, p. 4.

2 Len Fisher, "Equations for Everyday Living," *New Scientist*, July 30, 2005; Simon Singh, "Lies, Damn Lies and PR," *New Scientist*, August 20, 2005.

3 For a discussion of this point, see William Steinhoff, *George Orwell and the Origins of* 1984 (Ann Arbor: University of Michigan Press, 1975), chapter XII.

4 Eugene Lyons, writing about the Soviet Union's first Five Year Plan,

quoted in Steinhoff, *George Orwell and the Origins of* 1984, p. 172.

5 Quoted in Robert A. Orsi, "2 + 2 = 5," *American Scholar* 76 (Spring 2007), pp. 34–43.

CHAPTER FIVE The Scientific Equivalent of Shakespeare

1 Maxwell to Lord Rayleigh, 1870. James Clerk Maxwell, *The Scientific Letters and Papers of James Clerk Maxwell, Vol. II: 1862–1873* (Cambridge: Cambridge University Press, 1995), p. 583.

2 Wilhelm Wien, "A New Relationship Between the Radiation from a Black Body and the Second Law of Thermodynamics," in *Sitzungsberichte der Königlich Preussischen Akademie der Wissenschaften zu Berlin*, 1893 pp. 55–62 at p. 62.

3 Max Planck, "On an Improvement of the Wien's Law of Radiation," *Verhandl. Dtsch. Phys. Ges.* 2 (1900), p. 202.

4 Kelvin, "Nineteenth Century Clouds over the Dynamical Theory of Heat and Light," in *Baltimore Lectures on Molecular Dynamics and the Wave Theory of Light* (London: Cambridge University Press, 1904), pp. 486–527.

CHAPTER SIX "The Most Significant Event of the 19th Century"

1 P. M. Harman, ed., *The Scientific Letters and Papers of James Clerk Maxwell*, vol. 1 (Cambridge: Cambridge University Press, 1990), p. 254.

2 James Clerk Maxwell, *A Treatise on Electricity and Magnetism* (New York: Dover, 1954), p. ix.

3 William Thomson, *Kelvin's Baltimore Lectures and Modern Theoretical Physics*, ed. R. H. Kargon and P. Achinstein (Cambridge: MIT Press, 1987), p. 206.

4 J. C. Maxwell, "Essay for the Apostles on 'Analogies in Nature,' " in *The Scientific Letters and Papers of James Clerk Maxwell*, vol. 1, ed. P. M. Harman (Cambridge: Cambridge University Press, 1990), pp. 376–83.

5 "On Faraday's Lines of Force," in *The Scientific Papers of James Clerk Maxwell*, vol. 1, ed. W. D. Niven (New York: Dover, 1965), pp. 155–229.

6 Maxwell, *The Scientific Papers*, p. 207.

7 In a Letter from M. Faraday to J. Maxwell, March 25, 1857, cited in Maxwell, *Scientific Letters and Papers*, p. 548.

8 "On Physical Lines of Force," in *The Scientific Papers*, p. 500.
9 Maxwell, *The Scientific Papers*, p. 533.
10 In 1868, Maxwell wrote a short paper, "A Note on the Electromagnetic Theory of Light" (*Scientific Papers* II, pp. 137–43), in which he admits that in his previous work on electromagnetic phenomena the connection to light was "not easily understood when taken by itself," and he restates the connection in "the simplest form," in the form of four theorems—but these are not yet "Maxwell's equations."
11 Cited in Dorothy M. Livingston, *The Master of Light* (New York: Scribner's 1973), p. 100.
12 J. Clerk Maxwell, "On a Possible Mode of Detecting a Motion of the Solar System through the Luminiferous Ether," *Nature* 21, January 29, 1880, pp. 314–15.
13 A good account of this meeting is given in B. J. Hunt, *The Maxwellians* (Ithaca: Cornell University Press, 1991), chapter 7.
14 Quoted in E. T. Bell, *Men of Mathematics* (New York: Simon and Schuster, 1937), p. 16.
15 Albert A. Michelson, "The Relative Motion of the Earth and the Luminiferous Ether," *American Journal of Science* 22 (1881), p. 120.
16 Quoted in Livingston, *Master of Light*, p. 77.
17 D.S.L. Cardwell, *The Organization of Science in England* (London: Heinemann), p. 124n.
18 Of special importance was the flux theorem. "At a time when work on Maxwell's theory could easily have wandered off into purely mathematical elaborations, the discovery of the energy flux theorem focused attention firmly on the physical state of the field." Hunt, *The Maxwellians*, p. 109.
19 Oliver Heaviside, *Electromagnetic Theory*, vol. 1 (New York: Chelsea, 1971), p. vii.
20 Oliver Heaviside, *Electrical Papers*, vol. 2 (New York: Chelsea, 1970), p. 525.
21 Hunt, *The Maxwellians*, p. 122.
22 Those are taken from the Appendix in Hunt's *The Maxwellians*, "From Maxwell's Equations to 'Maxwell's Equations,' " p. 247.
23 Heaviside, "On the Metaphysical Nature of the Propagation of Potentials," *Electrical Papers*, vol. 2, pp. 483–85.
24 Hunt, *The Maxwellians*, p. 128.

CHAPTER SEVEN Celebrity Equation

1 Dalai Lama, *The Universe in a Single Atom: The Convergence of Science and Spirituality* (New York: Morgan Road, 2005), p. 59.
2 Luce Irigaray, *Parler n'est jamais neutre* (Paris: Editions de Minuit, 1987), p. 110.
3 For a brief discussion of some Maxwell "modifiers," see Alfred Bork, "Physics Just Before Einstein," *Science* 152 (1966), pp. 597–603.
4 G. FitzGerald, "The Ether and the Earth's Atmosphere," *Science* 13 (1889), p. 390.
5 Lorentz to Rayleigh, August 18, 1892, cited in John S. Rigden, *Einstein 1905: The Standard of Greatness* (Cambridge, MA: Harvard University Press, 2005), p. 82.
6 G. FitzGerald to H. Lorentz, November 14, 1894, quoted in Abraham Pais, *"Subtle Is the Lord": The Science and Life of Albert Einstein* (New York: Oxford, 1982), p. 124.
7 One physicist I know remembers that time dilates in a rest frame by thinking the following: "Cosmic rays reach earth." In the rest frame, that is, cosmic rays have a lifetime that ordinarily is too short for them to travel long distances. But because from earth's point of view they are moving at speeds close to the speed of light, time is dilated for them long enough for them to reach the ground.
8 Arthur Eddington, "Gravitation and the Principle of Relativity," *Nature*, vol. 101, 1918, pp. 15–17 (quote appears on p. 16).
9 Pais, *"Subtle Is the Lord,"* p. 128.
10 Carl Seeling to Einstein, March 11, 1952. Quoted in Ronald W. Clark, *Einstein: The Life and Times* (New York: World Publishing, 1971), p. 84.
11 P. A. Schilpp, ed., *Albert Einstein: Philosopher-Scientist* (London: Cambridge University Press, 1970), p. 53.
12 Quoted in Clark, *Einstein*, p. 84.
13 Quoted in Pais, *"Subtle Is the Lord,"* p. 139.
14 Emilio Segre, *From X-rays to Quarks* (New York: Dover, 1980), p. 84.
15 A. Einstein, "On the Electrodynamics of Moving Bodies," *Annalen der Physik* 17 (1905), pp. 891–921, in Albert Einstein, *The Collected Papers of Albert Einstein, Vol. 2. The Swiss Years: Writings, 1900–1909,* trans. A. Beck (Princeton: Princeton University Press, 1989), doc. 23, pp. 140–71.
16 Einstein to Conrad Habicht, June 30, 1905. In *Collected Papers,* vol. 5, pp. 20–21.
17 Rigden, *Einstein 1905*, p. 112.
18 Einstein, *Collected Papers*, vol. 2, doc. 24, p. 174.

19 Abraham Pais writes of Einstein's achievement: "In physics the great novelties were, first, that the recording of measurements of space intervals and time durations demanded more detailed specifications than were held necessary theretofore and, second, that the lessons of classical physics are correct only in the limit $v/c \ll 1$. In chemistry the great novelty was that Lavoisier's law of mass conservation and Dalton's rule of simply proportionate weights were only approximate but nevertheless so good that no perceptible changes in conventional chemistry were called for. Thus relativity turned Newtonian mechanics and classical chemistry into approximate sciences, not diminished but better defined in the process." Pais, "*Subtle Is the Lord*," p. 163.

20 This is, again, a requirement of objectivity or covariance that drives a new wedge between ordinary notions of objectivity and scientific ones.

21 A. Einstein, "The Principle of Conservation of Motion of the Center of Gravity and the Inertia of Energy," *Annalen der Physik* 20 (1906), pp. 627–33, in *Collected Works*, vol. 2, pp. 200–206.

22 A. Einstein, "On the Inertia of Energy Required by the Relativity Principle," *Annalen der Physik* 23 (1907), pp. 371–84, in *Collected Works*, vol. 2, p. 249.

23 A. Einstein, "On the Relativity Principle and the Conclusions Drawn from it," *Collected Works*, vol. 2, pp. 286–87.

24 A. Einstein, *Einstein's 1912 Manuscript on the Special Theory of Relativity* (New York: Braziller, 1996), pp. 102–3, 109.

25 "A. Einstein, $E = mc^2$: The Most Urgent Problem of Our Time," *Science Illustrated* (April 1946), pp. 16–17.

26 M. Planck, quoted in Einstein, *Collected Works*, vol. 2, p. 287.

27 Clark, *Einstein*, p. 101.

28 Niels Bohr, *Nature*, February 29, 1936.

29 Abraham Pais, *J. Robert Oppenheimer: A Life*, with supplemental material by Robert P. Crease (New York: Oxford, 2006), p. 44.

30 *The New York Times*, August 7, 1945, p. 1.

31 Henry D. Smyth, *Atomic Energy for Military Purposes: The Official Report on the Development of the Atomic Bomb under the Auspices of the United States Government, 1940–1945* (Princeton: Princeton University Press, 1945).

32 "A. Einstein, $E = mc^2$," *Science Illustrated.*

INTERLUDE Crazy Ideas

1 Jeremy Bernstein, *Science Observed* (New York: Basic Books, 1982), p. 310.
2 Bernstein once noted several characteristics of crackpots. They insist that their work has solved *everything*, they are humorless, they are sure everyone is out to steal their ideas, they are sure the media will be interested, they use a lot of capital letters. "Scientific Cranks," in *Science Observed*, ch. 14.

CHAPTER EIGHT The Golden Egg

1 *Times of London*, November 8, 1919, p. 1.
2 Quoted in Abraham Pais, *"Subtle Is the Lord": The Science and Life of Albert Einstein* (New York: Oxford, 1982), p. 124.
3 A. N. Whitehead, *Science and the Modern World* (New York: Macmillan, 1954), p. 13.
4 Quoted in Pais, *"Subtle Is the Lord,"* p. 179.
5 Ibid., p. 178.
6 A. Einstein, *The Collected Works of Albert Einstein*, vol. 2, trans. A. Beck (Princeton: Princeton University Press, 1989), pp. 301–2.
7 Ibid., p. 310.
8 A. Einstein to C. Habicht, December 24, 1907, in *Collected Works*, vol. 5, p. 47.
9 Quoted in Ronald W. Clark, *Einstein: The Life and Times* (New York: World Publishing, 1971), p. 120.
10 Emilio Segre, *From X-rays to Quarks* (New York: Dover, 1980), p. 85.
11 Quoted in Pais, *"Subtle Is the Lord,"* p. 152.
12 A. Einstein, "On the Influence of Gravitation on the Propagation of Light," *Annalen der Physik* 35 (1911), pp. 898–908, in *Collected Works,* vol. 3, p. 379.
13 A. Einstein to Willem Julius, August 24, 1911, in *Collected Works*, vol. 5, p. 199.
14 A. Einstein to E. Freundlich, September 1, 1911, in ibid., p. 202.
15 J. Earman and C. Glymour, "Relativity and Eclipses: The British Eclipse Expeditions of 1919 and their Predecessors," *Historical Studies in the Physical Sciences* 11 (1980), p. 61.
16 A. Einstein to E. Mach, June 25, 1913, in *Collected Works*, vol. 5, p. 340.
17 Quoted in Pais, *"Subtle Is the Lord,"* p. 311.
18 Ibid., p. 212.

19 A. Einstein to L. Hopf, August 16, 1912, in *Collected Works*, vol. 5, p. 321.

20 A. Einstein to A. Sommerfeld, October 29, 1912, in *Collected Works*, vol. 5, p. 324.

21 *Zeitschrift für Mathematik und Physik* 62 (1913), pp. 225–61. The "hair's breadth" remark is from John Norton, "How Einstein Found His Field Equations: 1912–1915," *Historical Studies in the Physical Sciences* 14:2 (1984), pp. 253–316.

22 A. Einstein to H. Lorentz, August 16, 1913, in *Collected Works*, vol. 5, p. 352.

23 Quoted in Clark, *Einstein*, p. 173.

24 Ibid., p. 199.

25 A. Einstein to A. Sommerfeld, November 28, 1915, in *Collected Works*, vol. 8, p. 152.

26 Pais, "*Subtle Is the Lord*," p. 253.

27 A. Einstein to H. Lorentz, January 16, 1915, in *Collected Works*, vol. 8, p. 179.

28 Quoted in Pais, "*Subtle Is the Lord*," p. 253.

29 A. Einstein, "Explanation of the Perihelion Motion of Mercury from the General Theory of Relativity," November 18, 1915, in *Collected Works*, vol. 6, p. 113.

30 Ibid., p. 117.

31 Independently, mathematician David Hilbert produced a similar equation.

32 Quoted in Clark, *Einstein*, p. 200.

33 A. Einstein to H. Lorentz, January 17, 1916, in *Collected Works*, vol. 8, p. 179.

34 And the following year, in a paper called "Cosmological Considerations in the General Theory of Relativity," Einstein tinkered with his basic field equation. He had noted that it seemed to suggest that the universe is expanding, so he subtracted from the left-hand side of the equation ($G_{\mu\nu}$) another tensor $g_{\mu\nu}$, multiplied by a constant λ, whose value, he admitted, was "at present unknown." The result kept the general covariance, as well as what he evidently assumed was a finite universe. This turned his field equation

$$G_{\mu\nu} = -\kappa(T_{\mu\nu} - \tfrac{1}{2}g_{\mu\nu}T)$$

into

$$G_{\mu\nu} - \lambda\, g_{\mu\nu} = -\kappa(T_{\mu\nu} - \tfrac{1}{2}g_{\mu\nu}T)$$

Einstein introduced this factor—the now-famous cosmological constant—purely as a fudge factor, to save what he thought was a prediction of his theory that the universe was expanding. Within a few years he would begin to question the necessity of this concept, and in 1931 removed the constant λ from the theory for good, later calling this fudge the "biggest blunder" of his life. Seventy years later, to explain data from measurements of supernovae, astronomers restored it.

35 See's article, "Einstein a Trickster," is reproduced in Jeffrey Crelinsten, *Einstein's Jury: The Race to Test Relativity* (Princeton: Princeton University Press, 2006), p. 222.

36 The classic article on the eclipse is J. Earman and Clark Glymour, "Relativity and eclipses: The British eclipse expeditions of 1919 and their predecessors," *Historical Studies in the Physical Sciences* 11:1 (1980), pp. 49–85.

37 Alistair Sponsel, "Constructing a 'Revolution in Science': The Campaign to Promote a Favorable Reception for the 1919 Solar Eclipse Experiments," *British Journal For the History of Science* 35 (2002), pp. 439–68.

38 A. Einstein to Pauline Einstein, September 27, 1919, in *Collected Works*, vol. 9, p. 98.

39 *Naturwissenschaften* 7 (1919), p. 776.

40 Quoted in Clark, *Einstein*, p. 230.

41 "Joint Eclipse Meeting of the Royal Society and the Royal Astronomical Society," *The Observatory* 42 (November 1919), p. 389.

42 Eddington *Relativity*, Eighth Annual Haldane Lecture may 26, 1937.

43 Albert Einstein, *Ideas and Opinions* (New York: Bonanza Books, 1954), p. 311.

CHAPTER NINE "The Basic Equation of Quantum Theory"

1 W. Nernst, quoted in M. Jammer, *The Conceptual Development of Quantum Mechanics* (New York: McGraw Hill, 1966), p. 59.

2 Jammer, *The Conceptual Development*, p. 170.

3 Ibid., p. 178.

4 "It appears to me that hydrogen," Balmer wrote prophetically in his paper, "more than any other substance is destined to open new paths to the knowledge of the structure of matter and its properties." Quoted in Jammer, *The Conceptual Development*, p. 65.

5 A. Einstein to C. Habicht, May 1905, in *Collected Works*, vol. 5, p. 20.

6 A. Einstein, "On the Quantum Theory on Radiation," in *Collected Works*, vol. 6, pp. 220–33.
7 Ibid.
8 *Physical Review* 21 (1923), 483–502.
9 Jammer, *The Conceptual Development*, p. 171.
10 N. Bohr, H. Kramers, and J. Slater, "The Quantum Theory of Radiation," *Philosophical Magazine* 47 (1924), p. 785.
11 Jammer, *The Conceptual Development*, p. 196.
12 The remarkable route to the wave equation has been extensively analyzed by several historians of science, including Martin Klein, "Einstein and the Wave-Particle Duality," *The Natural Philosopher* 3 (1964), pp. 3–49; L. Wessels, "Schrödinger's Route to Wave Mechanics," *Stud Hist Phil sci* 10 (1979), pp. 311–40; M. Jammer, *The Conceptual Development of Quantum Mechanics*, 1966, ch. 5, sec. 3.
13 Walter Moore, *A Life of Erwin Schrödinger* (Cambridge: Cambridge University Press, 1994), pp. 195–96.
14 Quoted in Mara Beller, *The Genesis of Interpretations of Quantum Mechanics 1925–1927*, PhD Dissertation, University of Maryland, 1983, p. 124.
15 Moore, *A Life*, pp. 195–96.
16 Quoted in ibid.
17 Jammer, *The Conceptual Development,* p. 267.
18 E. Schrödinger, *Collected Papers on Wave Mechanics* (Providence, RI: AMS/Chelsea Publishing, 1982), p. 59.
19 Ibid., p. 20.
20 Ibid., p. 9.
21 Jammer, *The Conceptual Development*, p. 284.
22 Born, "Physical Aspects of Quantum Mechanics," *Nature* 119 (1926), pp. 354–57.
23 Quoted in Jammer, *The Conceptual Development*, p. 285.
24 Quoted in Beller, *Genesis*, p. 144.
25 Beller, *Genesis*, p. 105.

INTERLUDE The Double Consciousness of Scientists

1 Quoted in Mara Beller, *The Genesis of Interpretations of Quantum Mechanics 1925–1927*, PhD Dissertation, University of Maryland, 1983, p. 86.
2 Ibid., p. 91.

CHAPTER TEN Living with Uncertainty

1 Anne Bogart and Kristin Linklater, "Balancing Acts," *American Theatre*, January 2001.
2 David Cassidy, *Uncertainty: The Life and Science of Werner Heisenberg* (New York: Freeman, 1992).
3 Interview, Werner Heisenberg, February 27, 1963, *Archives for the History of Quantum Physics* [hereafter *AHQP*] (College Park, MD: American Institute of Physics), p. 22. This is not what Heisenberg would have said at the time. At that time, he was trying to get rid of classical notions altogether.
4 W. Heisenberg, "Erinnerungen an der Zeit die Entwicklung der Quantenmechanik," in *Theoretical Physics in the Twentieth Century*, ed. M. Fierz and V. F. Weisskopf (New York: Interscience, 1960).
5 Isn't it possible, indeed, that the human perceptual and imaginative capacities have evolved to handle environments of a scale about that of the human body, rather than for the microworld, a billion orders of magnitude smaller?
6 Mara Beller, *Quantum Dialogue: The Making of a Revolution* (Chicago: University of Chicago Press, 1999), p. 22.
7 Patrick A. Heelan, *Quantum Mechanics and Objectivity* (The Hague: Nijhoff, 1965), p. 23.
8 Max Born, *My Life and My Views* (New York: Scribner's, 1968), p. 216.
9 W. Heisenberg, *Physics and Beyond: Encounters and Conversations* (New York: Harper and Row, 1971), p. 60.
10 Ibid., p. 61.
11 W. Heisenberg, "On the Quantum-Mechanical Reinterpretation of Kinematic and Mechanical Relations," *Zeitschrift für Physik* (*ZfP*) 33 (1925), 879–93; in B. L. van der Waerden, *Sources of Quantum Mechanics* (Amsterdam: North-Holland, 1967), p. 261.
12 W. Heisenberg, *AHQP* Interview, February 15, 1963.
13 "Über quantentheoretische Umdeutung kinematischer und mechanisher Beziehungen," in *ZfP* 53 (1925), p. 893.
14 Quoted in Beller, *Quantum Dialogue*, p. 43.
15 M. Born, "Remarks at Le Banquet Nobel," in *Les Prix Nobel en 1954* (Stockholm: Royale P. A. Norstedt & Stoner, 1955).
16 Max Born, *Physics in My Generation* (New York: Pergamon Press, 1969), p. 100.
17 Nancy Greenspan, *The End of the Certain World: The Life and Science of Max Born* (New York: Basic Books, 2005), p. 127.
18 Before it was published, they received a shock, in the form of a paper

by a Cambridge student named Paul Dirac, who had been given a copy of Heisenberg's paper in Cambridge, studied it, and had come to the same conclusions as Born and Jordan, using slightly different language. Meanwhile, Dirac heard Heisenberg's talk, applied new notation, and came up with a distinction between **q** numbers and **p** numbers. The variables were not classical variables, or those satisfying the commutative law (what Dirac would soon call c-numbers), but symbols referring to quantum mechanical variables (q-numbers).

19 Quoted in Abraham Pais, *Inward Bound* (New York: Oxford University Press, 1986), p. 258.

20 One irony, pointed out by Jammer (*The Conceptual Development* p. 215), is that the mathematics of this attempt to rid atomic physics of a solar-system-like picture was based on so-called secular equations that derived from methods of astronomers to compute planetary orbits.

21 Quoted in Beller, *Genesis*, p. 81.

22 E. Schrödinger, "The Continuous Transition from Micro- to Macro-Mechanics," *Die Naturwissenschaften* 28 (1926), pp. 664–66, in Schrödinger, *Collected Papers*, pp. 41–44.

23 Jammer, *The Conceptual Development*, p. 271.

24 Quoted in Beller, *Genesis*, pp. 85, 89.

25 M. Born to E. Schrödinger, November 6, 1926, in *AHQP*.

26 Schrödinger, *Collected Papers*, pp. 45–61.

27 Beller, *Genesis*, p. 93.

28 Cassidy, *Uncertainty*, p. 215.

29 Schrödinger, *Collected Papers*, p. 46.

30 Ibid., p. 59.

31 E. Schrödinger to W. Wien, June 1926, cited in Cassidy, *Uncertainty*, p. 214.

32 Beller, *Genesis*, p. 207.

33 Beller, *Quantum Dialogue*, p. 410.

34 W. Heisenberg, *Physics and Beyond*, p. 73.

35 Pauli put this interpretation in a footnote to one of his papers: "Über Gasentartung und Paramagnetismus," *ZfP* 41, 1927.

36 Quoted in Beller, *Genesis*, p. 137.

37 W. Pauli to W. Heisenberg, October 19, 1926, in A. Hermann, K. Meyenn, and V. Weisskopf, *Wolfgang Pauli: Wissentschaftlicher Briefwechsel mit Bohr, Einstein, Heisenberg, u.a.* (New York: Springer, 1979), p. 347.

38 W. Heisenberg to W. Pauli, November 15, 1926, ibid., p. 355.

39 W. Heisenberg to W. Pauli, November 23, 1926, ibid., p. 359.

40 In effect, Jordan's article implicitly expresses what many people are tempted to think when they first encounter the uncertainty principle—that electrons and other tiny bits of matter *really do* have positions and momenta, but that we cannot track them down because of some defect—maybe even an essential and ineradicable defect—in our measuring instruments.

41 John H. Marburger, III, "A Historical Derivation of Heisenberg's Uncertainty Relation Is Flawed," *American Journal of Physics* 76 (2008), pp. 585–87.

42 Beller, *Genesis*, p. 217.

43 Quoted in ibid., p. 318.

44 W. Heisenberg, *AHQP* Interview, February 25, 1963.

45 Beller, *Genesis*, p. 245ff.

46 N. Bohr, *Atomic Theory and the Description of Nature* (Cambridge: Cambridge University Press, 1934), p. 54.

47 Quoted in "The Philosophy of Niels Bohr," by Aage Peterson, in *Bulletin of the Atomic Scientists* 19, no. 7 (1963).

48 Quoted in Beller, *Genesis*, p. 248.

INTERLUDE The Yogi and the Quantum

1 P. W. Bridgman, "The New Vision of Science," *Harper's*, March 1929, pp. 443–51.

2 I am greatly indebted to John H. Marburger, III, for pointing this out to me. "It's a clear, logical, and consistent way of framing the complementarity issue," Marburger says. "It clarifies how quantum phenomena are represented in alternative classical 'pictures,' and it fits in *beautifully* with the rest of physics. The clarity of this scheme removes much of the mysticism surrounding complementarity. What happened was like a Gestalt-switch, from a struggle to view microscopic nature from a classical point of view to an acceptance of the Hilbert space picture, from which classical concepts emerged naturally. Bohr brokered that transition."

CONCLUSION Bringing the Strange Home

1 Peter Galison, *Image and Logic: A Material Culture of Microphysics* (Chicago: University of Chicago Press, 1997), p. 801.

2 Leon Lederman, "The Pleasure of Learning," *Nature* 430:5 (August 2004), p. 617.

ILLUSTRATION CREDITS

Excerpts from the column, "Critical Point," by Robert P. Crease, are reprinted from *Physics World* with the kind permission of the publisher.

Page 24: Columbia University.

Pages 25, 28, 30, 39, 73, 99, 101, 104, 201: John McAusland.

Page 43: Robert P. Crease.

Page 58: AIP Galilei Galileo A9.

Page 59: AIP Newton Isaac A6.

Page 83: AIP Newton Isaac H5.

Page 94: AIP Euler A1.

Page 112: Boltzmann Ludwig A4; AIP Clausius Rudolf A3; AIP Helmoltz Hermann A2; AIP Carnot Sadi A1; AIP Joule James A3; AIP Maxwell James Cleak A5; AIP Rumford Benjamin H2; AIP Emilio Segre Visual Archives, E. Scott Barr Collection; AIP Kelvin William Thomson A16; AIP Planck Max A14; Wien, Wilhelm A1.

Page 134: AIP Maxwell James Clerk A5.

Page 140: Maxwell, James Clerk. *A Treatise on Electricity and Magnetism, vol. 1*. Oxford: 1873.

Page 157: Sidney Harris.

Page 167: *Physics Today*, January 2006.

Page 179: *Time* magazine.

Page 221: AIP Schrodinger, Erwin A10.

Page 237: Heisenberg Werner A15.

ACKNOWLEDGMENTS

This book, like my previous book, *The Prism and the Pendulum: The Ten Most Beautiful Experiments in Science*, grew out of a column I wrote for *Physics World.* Once again I am grateful to its editors, especially Matin Durrani, for allowing me to write a column for that magazine, as well as to the hundreds of people who responded to my column about great equations. That column made it possible for me to try out many ideas in this book, and bits and pieces of my columns show up throughout. I am indebted to my literary agent, John Michel, for steering me again and again in the right direction; to Margaret Maloney, for helping me through the manuscript process; to production manager Julia Druskin; and to my editor, Maria Guarnaschelli, for her thoughtful reading and guidance. Like all columnists, I rely heavily on colleagues and correspondents for inspiration, ideas, and information, and those who provided helpful suggestions, comments, and other kinds of assistance include: Edward S. Casey, David Cassidy, Carlo Cercignani, Allegra de Laurentiis, John de Pillis, David Dilworth, B. Jeffrey Edwards, Elizabeth Garber, Patrick Grim, Richard Harrison, George W. Hart, Richard Howard, Don Ihde, Eric Jones, Ed Leibowitz, Gerald M. Lucas, Bob Lloyd, Peter Manchester, Eduardo Mendieta, Hal Metcalf, Lee Miller, Eli Maor, Anthony Phillips, Xi Ping, Mary Rawlinson, Robert C. Scharff, David Socher, Marshall Spector, Clifford Swartz, Dick Teresi, Beth Young. Without the capable help of Alissa Betz, Ann-Marie Monaghan, and Nathan Leoce-Schappin in the Department of Philosophy office this manuscript would have been much delayed. John H. Marburger, III, helped me to avoid numerous errors of fact and interpretation in the chapters on quantum mechanics, though I'm sure I still managed to commit some. Alfred S. Goldhaber provided me with thoughtful advice on the quantum mechanics and relativity chapters, and I benefited from many discussions that I had with him while co-teaching a course on the influence of quantum mechanics outside physics. My wife, Stephanie, not only read the manuscript but put up with my difficult—some would say impossible—work routines, which are trying to anyone within range, but nevertheless managed to provide me throughout with the sounds of surprise. My son, Alexander, likewise had to endure my work habits and periodic unavailability, and drew several of the diagrams. My daughter, India, always made sure I was in the right dimension. My dog, Kendall, was always willing to go for a walk with me when the rest of them got fed up. And again I want to thank Charles C. Mann, the finest science writer of this generation, for his generosity, inspiration, and example.

INDEX

Page numbers in *italics* refer to illustrations.

FASKEN LEARNING RESOURCE CENTER
9000087496

About the Author

Robert P. Crease is a professor in and chairman of the Department of Philosophy at Stony Brook University in New York, and historian at Brookhaven National Laboratory. He writes a monthly column, "Critical Point," for *Physics World* magazine. His previous books include *The Prism and the Pendulum: The Ten Most Beautiful Experiments in Science*; *Making Physics: A Biography of Brookhaven National Laboratory*; *The Play of Nature: Experimentation as Performance*; *The Second Creation: Makers of the Revolution in Twentieth-Century Physics* (with Charles C. Mann); and *Peace & War: Reminiscences of a Life on the Frontiers of Science* (with Robert Serber). Crease's translations include *American Philosophy of Technology: The Empirical Turn* and *What Things Do: Philosophical Reflections on Technology, Agency, and Design*. He lectures widely, and his articles and reviews have appeared in *The Atlantic Monthly*, *The New York Times Magazine*, *The Wall Street Journal*, *Science*, *New Scientist*, *American Scientist*, *Smithsonian*, and elsewhere. He lives in New York City.